Methods of Soil and Water Analysis

Methods of Soil and Water Analysis

Authors

Shashi S. Yadav

S.K. Verma

Priyadarshani A. Khambalkar

Akhilesh Singh

SATISH SERIAL PUBLISHING HOUSE

403, Express Tower, Commercial Complex
Azadpur, Delhi-110033 (India)
Phone : 011-27672852, Fax : 91-11-27672046
e-mail : info@satishserial.com, hkjain1975@yahoo.com
Website : www.satishserial.com

Published by :

SATISH SERIAL PUBLISHING HOUSE

403, Express Tower, Commercial Complex, Azadpur, Delhi-110033 (INDIA)
Phone : 011-27672852 Fax : 91-11-27672046
E-mail : info@satishserial.com, hkjain1975@yahoo.com

ISBN 978-93-88020-71-8

Composed, Designed & Printed in India

About the Authors

Dr. (Smt.) Shashi S. Yadav is a permanent resident of village Mangrouth, Post Office Gohand via Rath, District Hamirpur (U.P.) is working as Scientist, Soil Science at RVSKVV, College of Agriculture, Gwalior (M.P.). She is graduated from Agriculture College, Nagpur, Dr. P.D.K.V. Akola. Thereafter she did her M.Sc. (Ag.) in Soil Science with specialization in land resource management from Dr. P.D.V.K. Akola in collaboration with ICAR-NBSS & LUP, Nagpur, Maharashtra and Ph.D. in Soil Science from MPUA & T, Udaipur. She has received ICAR JRF for M.Sc. (Ag.) and SRF for Ph.D. (Ag.) in Soil Science, passed ICAR-NET in Soil Science-Pedology in 1997. She is recipient of Young Scientist award in 2016 and Best Teacher award in 2018 by academy for Environment & Life Science. Her area of interest of research is Soil Science-Pedology. She has published more than 20 research papers in NAAS rated journals, 8 technical bulletins and 6 Chapter in edited books.

Dr. Surendra Kumar Verma born in 1954 and graduated in science discipline from Jiwaji University, Gwalior in the year 1974. He completed his post graduation in Soil Science & Agricultural Chemistry from JNKVV, Jabalpur in the year 1977. After post graduation he joined services as Jr. Scientist in the year 1978 in all India coordinated project on soil salinity and water management. After completing five years of his services he was allowed for joining as Ph.D. scholar from same university and completed degree in 1988. He has been Asstt. Professor/Jr. Scientist from 1978 to 1986 and Scientist/Associate Professor, 1986-2004 and got promoted as Professor/Principal Scientist in the year 2004. He has got published teaching/research experience of more than 40 years and published over 100 research papers. He has another of several chapters and books related to soil science.

Dr. Priyadarshani A. Khambalkar is presently working as Assistant Professor (SSAC) at RVSKVV, CoA, Gwalior, M.P. She is graduated from Anandan Niketan College of Agriculture, Warora, Dr. PDKV, Akola, Maharashtra. She did her M.Sc. (Ag.) in SSAC with specialization in LRM from Dr. PDKV Akola in association with NBSS & LUP. She completed her Ph.D. (SSAC) and awarded by Gold medal (RVSKVV, CoA, Gwalior, M.P.). She qualified ASRB- NET and appeared for interview ICAR-ARS, (2011). She is recipient of young Scientist Award of Society for Scientific and Social Development (2017). She has been awarded by Excellent teaching award (2019) by Academy for Environment & Life Sciences.

Dr. Akhilesh Singh born in 1975. He did B.Tech in Agricultural Engineering in 1999 from Allahabad Agricultural Institute, Allahabad and completed M.Tech. and Ph.D. in Soil and Water Engineering in 2002 and 2009, respectively from JNKVV, Jabalpur. He worked as SRF and RA for more than 7 years at College of Agricultural Engineering, JNKVV, Jabalpur. Presently he is working as a Scientist (Agril. Engg.) at RVSKVV, Gwalior since 2011. He has published more than 15 research papers in NAAS rated journal, 12 Technical bulletins/manuals/ practical manuals, one ISBN edited book too.

Preface

Soil and water are the major ingredients of interest in agriculture system. Analysis of soil and water as natural resources is of utmost important with regard to conservation strategies. About one-fourth of the land surface is in arid and semi-arid regions. Soils of these regions are generally characterized by their slightly alkaline reaction (pH 7.8-8.6), and elevated accumulation of calcium and magnesium-carbonates, and sulphates. Because of low rainfall, high evaporation, and restricted leaching, soluble salts accumulate to some high levels in certain areas, which lead to saline conditions. Abundance of lime and gypsum, and the accumulation of soluble salts, markedly influence the physical and chemical properties and the fertility status of these soils. The surface layers of most of these soils, which contain various kinds of deposits, are coarse textured and low in organic matters. These factors make the management of these soils difficult, and demand studies based on rightly adapted methods of soil analysis. The saying "test and don't guess" is appropriate for the management of these soils, and the determination of the problems related to irrigation and drainage, salinity, sodicity, and gypsum content. A number of books are available describing numerous alternative procedures for soil analysis, and users are often confused in selecting the appropriate methods. This publication aims to establish a middle ground between the so-called "cook-book" and the detailed comprehensive type of manual. In doing so, analytical methods, which have been developed and adapted specially to the particular conditions of these soils, are described. It is believed that this publication will be of particular interest to students, teachers, researchers and chemists working in laboratories for providing soil testing and analytical services to farmers in arid and semi-arid regions. It is also hoped that this manual will be a valuable reference to organizations helping in agricultural development projects in these regions.

The manual provides elaborated information on major soil types of India, their composition, apart from procedure of sample collection and methods of analysis. Much of the information has been gathered from internet and books.

Contents

Chapter 1

General Laboratory Guidelines

In chemical laboratories, the use of acids, alkalis and some hazardous and explosive chemicals is inescapable. Apart from this, some chemical reactions during the process of analysis may release toxic gases and if not handled well, may cause explosion. Inflammable gases are also used as a fuel/heating source. Thus, safe working in a chemical laboratory needs special care, both in terms of design and construction of the laboratory building, and handling and use of chemicals. For chemical operations special chambers also need to be provided.

Air temperature of the laboratory and working rooms should be maintained at a constant level between 20-25°C. Humidity should be kept at about 50%. Soil samples are often affected by the temperature and humidity. The temperature influences even some chemical operations. Hence, maintenance of temperature and humidity as specified is critical. Proper air circulation is also important so as hazardous and toxic fumes and gases do not stay in the laboratory for long. The release of gases and fumes in some specific analytical operation are controlled through fume hood, trapped in acidic/ alkaline solutions and washed through flowing water. Maintenance of clean and hygienic environment in the laboratory is essential for the good health of the workers.

Care is needed to be taken to store acids and hazardous chemicals in separate and safe racks. An inventory of all the equipment, chemicals, glassware and miscellaneous items in a laboratory should be maintained. A safe laboratory building should have suitable separate rooms for different purposes and for performing different operations.

Laboratory Safety Measures

Special Care is required while operating equipment, handling the chemicals and in waste disposal.

Equipment

Electrical cables, plugs and tubing need proper check to avoid accident. Various types of gas cylinders needed in the laboratory like acetylene, nitrous oxide and LPG may be kept under watch and properly sealed/capped and may be stored in ventilated cupboards.

Chemical Reagents

Hazardous chemicals may be stored in plastic bottles. While working with chemicals, such as perchloric acid, fume hood may be used. Chemicals may be properly labelled indicating their hazardous nature.

Bottles with inflammable substances need to be stored in stainless steel containers.

Waste Disposal

Each country has special rules and methods for disposal of hazardous waste. Cyanides, chromates, arsenic, selenium, cobalt and molybdate are very commonly used but hazardous chemicals and should never be disposed off in the laboratory sink but collected in a metal container for proper disposal at the specified places and in the manner as described in the country's law for waste disposal.

General Rules and Required Care

- Learn safety rules and use of first aid kits. Keep the first aid kit handy at a conspicuous working place in the laboratory.
- Personal safety aids, such as laboratory coat, hand protection gloves, safety glasses, face shield and proper footwear should be used while working in the laboratory.
- Observe normal laboratory safety practice in connecting equipment with power supply, in handling chemicals and preparing solutions of reagents. All electrical work must be done by qualified personnel.

- Maintain instrument manual and logbook of reach equipment to avoid mishandling, accident and damage to equipment.
- Keep the working tables/space clean. Clean up spillage immediately.
- Wash hands after handling toxic/hazardous chemical.
- Never suck the chemicals with mouth but use automatic pipetting device.
- Use forceps/tongs to remove containers from the hot plates/ovens/furnaces.
- Do not use laboratory glassware for eating/drinking.
- Use fume hood while handling concentrated acids, bases and hazardous chemicals.
- Never open a centrifuge cover until the machine has stopped.
- Add acid to water and not water to acid while diluting the acid.
- Always put labels on bottles, vessels and wash bottles containing reagents, solutions, samples and water.
- Handle perchloric acid in fume hoods. Avoid direct contact with organic matter/rubber. In wet oxidation method of sample digestion, destroy organic matter first with nitric acid.
- Read the labels of the bottles before opening them.

❑❑❑

Chapter 2

Preparation and Standardization of Reagent Solutions

Chemical reagents are manufactured and marketed in different grades of purity.In general, the purest reagents are marketed as "analytical reagent" or AR-grade.Other labels are "LR", meaning laboratory reagent, and "CP", meaning chemically pure. The strength of chemicals is expressed as normality or molarity. Therefore, it is useful to have some information about the strength of the acids and alkalis most commonly used in chemical laboratories. Some important terms that are often used in a laboratory for chemical analysis are defined/ explained below.

Molarity

A one-molar (M) solution contains one mole or one molecular weight in grams of a substance in each litre of the solution. The molar method of expressing concentration is useful because equal volumes of equimolar solutions contain equal number of molecules.

Normality

The normality of a solution is the number of gram equivalents of the solute per litre of the solution. It is usually designated by the letter N. Semi-normal, penti-normal, desi-normal, centi-normal and milli-normal solutions are often required; these are written as 0.5N, 0.2N, 0.1N, 0.01N and 0.001N, respectively. However, molar expression is preferred because "odd" normalities, such as 0.121N are clumsily represented in fractional form. The definition of normal solution uses the term "equivalent weight". This quantity varies with the type of reaction. Hence, it is difficult to give a clear definition of equivalent weight that covers all reactions. It often happens that the same compound possesses different equivalent weights in different chemical reactions. A situation may arise where a solution has a normal concentration when employed for one purpose and a different normality when used in another chemical reaction. Hence, the system of molarity is preferred.

Equivalent weight

The equivalent weight (Eq W) of a substance is the weight in grams that in its reaction corresponds to a gram atom of hydrogen or of hydroxyl, or half a gram atom of oxygen, or a gram atom of univalent ion. When one equivalent weight of a substance is dissolved in one litre, it gives 1N solution.

Milli-equivalent weight

Equivalent weight when expressed as milli-equivalent weight (mEq W) means the equivalent weight in grams divided by 1 000. It is commonly expressed as "mEq". It is the most convenient value because it is the weight of a substance contained in or equivalent to 1 ml of 1N solution. Therefore, it is a unit that is common to both volumes and weights, making it possible to convert the volume of a solution to its equivalent weight, and the weight of a substance to its equivalent volume of solution : number of mEq = volume × normality.

Buffer solutions

Solutions containing either a weak acid and its salt or a weak base and its salt (e.g. $CH_3COOH + CH_3COONa$ and $NH_4OH + NH_4Cl$) possess the characteristic property of resisting changes in pH, when some acid or base is added to them. Such solutions are called buffer solutions. The important properties of a buffer solution are:

- It has a definite pH value.
- Its pH value does not alter on keeping for a long time.
- Its pH value is only altered slightly when a strong base or strong acid is added. Because of the above property, readily prepared buffer solutions of known pH are used in order to check the accuracy of pH metres used in the laboratory.

Titrations

Titration is a process of determining the volume of a substance required to just complete the reaction with a known amount of other substance. The solution of known strength used in the titration is called the titrant. The substance to be determined in the solution is called the titrate.The completion of the reaction is assessed with the help of an appropriate indicator.

Indicators

A substance that indicates the endpoint on completion of the reaction is called an indicator. The most commonly used indicators in volumetric analysis are:

- Internal indicators,
- External indicators,
- Self-indicators.

❑❑❑

Chapter 3

The Basics of An Analytical Laboratory

Internal indicators

Indicators, such as methyl red, methyl orange, phenolphthalein and diphenylamine that are added to the solution where reaction occurs are called internal indicators. On completion of the reaction of titrant on titrate, a colour change takes place owing to the presence of the indicator, which also helps in knowing that the titration is complete. The internal indicators used in acid-alkali neutralization solutions are methyl orange, phenolphthalein and bromothymol blue. The indicator used in precipitation reactions, such as titration of neutral solution of NaCl (or chlorine ion) with silver nitrate ($AgNO_3$) solution is K_2CrO_4 -. On completion of the titration reaction of $AgNO_3$ with Cl, when no more Cl is available for reaction with silver ions to form AgCl, the chromium ions combine with Ag^{2+} ions to form sparingly soluble Ag_2CrO_4, which is brick red in colour. This indicates that Cl has been titrated completely and that endpoint has occurred.Redox indicators are also commonly used. These are the substances that possess different colours in the oxidized and reduced forms. Diphenylamine has a blueviolet colour in the oxidation state and is colourless in reduced condition. Ferrocin gives a blue colour in the oxidation state and a red colour in reduced condition.

External indicators

Some indicators are used outside the titration mixture. Potassium ferricyanide is used as an external indicator in the titration of potassium dichromate and ferrous sulphate in acid medium. In this titration, a few drops of indicator are placed on a white porcelain tile. A glass rod dipped in the solution being titrated is taken out and brought into contact with the drops of indicator on the tile. First, a deep blue colour is noticed, which turns greenish on completion of titration.

Self-indicators

After completion of the reaction, the titrant leaves its own colour owing to its slight excess in minute quantities. In $KMnO_4$ titration with ferrous sulphate, the addition of $KMnO_4$ starts reacting with $FeSO_4$, which is colourless. On completion of titration, the slight excess presence of $KMnO_4$ gives a pink colour to the solution, which acts as a self-indicator and points to the completion of the titration.

Standard solutions

A solution of accurately known strength (or concentration) is called a standard solution. It contains a definite number of gram equivalent or gram mole per litre of solution. If it contains 1 g equivalent weight of a substance/compound, it is 1N solution. If it contains 2 g equivalent weight of the compound, it is 2N. All titrimetric methods depend on standard solutions that contain known amounts of the reagents in unit volume of the solution. A solution is prepared, having about the desired concentration. This solution is then standardized by *titrating* it with another substance that can be obtained in highly purified form. Thus, potassium permanganate solution can be standardized against sodium oxalate, which can be obtained in a high degree of purity as it is easily dried and is non-hygroscopic. Such a substance, whose weight and purity is stable, is called a "primary standard". A primary standard must have the following characteristics:

- It must be obtainable in a pure form or in a state of known purity.
- It must react in one way only under the condition of titration and there must be no side-reactions.
- It must be non-hygroscopic. Salt hydrates are generally not suitable as primary standards.
- Normally, it should have a large equivalent weight in order to reduce the error in weighing.
- An acid or a base should preferably be strong, that is, it should have a high dissociation constant for being used as standards.

A primary standard solution: It is one that can be prepared directly by weighing the material and with which other solutions of approximate strength can be titrated and standardized. Some primary standards are :

- Acids
  - Potassium hydrogen phthalate,
  - Benjoic acid;
- Bases
  - Sodium carbonate,
  - Borax;
- Oxidizing agents
  - Potassium dichromate,
  - Potassium bromate;
- Reducing agents
  - Sodium oxalate,
  - Potassium ferrocyanide;
- Others
  - Sodium chloride,
  - Potassium chloride.

Secondary standard solutions: They are those that are prepared by dissolving a little more than the gram equivalent weight of the substance per litre of the solution, and then their exact standardization is done with primary standard solution. Some secondary standards are:

- Acids
  - Sulphuric acid,
  - Hydrochloric acid;
- Base
  - Sodium hydroxide.

Standard solutions of all the reagents required in a laboratory must be prepared and kept ready before commencing any analysis. However, their strength should be checked periodically, or fresh reagents should be prepared before analysis.

All titrations involving acidimetry and alkalimetry require standard solutions.These may be prepared either from standard substances by direct weighing, or by standardizing a solution of approximate normality of materials by titrating against a prepared standard. The methods for preparing standard solutions of some non-primary standard substances in common use are given below.

□□□

Chapter 4

Quantitative Analysis

Quantitative analysis involves determination of the amount of element or radical present in the sample under examination: the amount is usually expressed either in gm/litre or-in terms of normality. This is done by either of the two processes: **(1) Gravimetric or (2) Volumetric.**

The substance unless supplied in solution, is brought into solution in an acid, a mixture of acids, an alkali, or in some other reagents. It is now reacted with a standard solution (i.e., a solution of accurately known concentration) of a suitable reagent; the reagent is carefully selected so that the reaction proceeds to completion and that the end of the reaction is indicated sharply either directly or by the addition of a suitable indicator. Then volume of the standard solution is carefully noted and from the volume the amount of the substance under investigation is calculated. This is called volumetric analysis.

1.1 Gravimetric analysis

The solution of the substance is treated with a suitable reagent under suitable conditions, so that the constituent under examination separates out completely as a precipitate in as pure a form as possible. The precipitate is then filtered out washed free from adhering chemicals, dried, ignited, if necessary, and weighed. From this weight the weight of the constituent under examination is calculated.

1.2 Volumetric analysis

In addition to the ordinary apparatus like beaker, funnel, etc, for volumetric work we require burette, pipette, and measuring flask of varying size. All apparatus should be cleaned carefully with distilled water.

$$\text{Strength of Borax solution} = \frac{wX4\ (N)}{190.72}$$

c) Standard Solution of Oxalic acid ($H_2C_2O_4$, $2H_2O$)

Equivalent weight of oxalic acid is 63.

Weigh out about 1.5 gram of oxalic acid (A. R.) into the funnel fitted to the neck of a 250 ml measuring flask. Determine the weight accurately. Let it be w.

Dissolve the acid as before in distilled water and make up the volume upto the mark. Shake well and allow to stand.

$$\text{Strength of oxalic acid soln.} = \frac{wX4\ (N)}{63}$$

Indicators: When an acid is titrated against an alkali solution or vice versa, it is necessary to determine the equivalence point or the end point very correctly. This is done with the help of an indicator (a neutralization or acid-base indicator). An indicator is a substance which show, one characteristic colour with an acidic solution and a different colour with an alkaline solution. The indicators mostly used are phenolphthalein, methyl orange and methyl red. Litmus is unsuitable for quantitative work.

Some indicators and their colours in acidic and alkaline media.

Indicators	Colour in acidic condition	Colour in alkaline
Phenolphthalein	Colourless	Red
Methyl orange	Red	Yellow
Methyl red	Red	Yellow
Bromocresol green	Yellow	Blue

Preparation of standard solution of Hydrochloric or Sulphuric acid.

Standard solutions of these acids can not be prepared by direct weighing. An approximately 0.1 *(N)* solution is first prepared and then its strength is accurately determined by titration against a standard solution of an alkali.

Concentrated hydrochloric acid is about 12 *(N)*

Concentrated sulphuric acid is about 36 *(N)*

Measure out by a graduated cylinder (or test tube), pipette, or burette 9 ml of concentrated hydrochloric acid and pour it into a litre flask containing 500 ml water, then add more of distilled water to make up the volume to the mark, shake and allow to stand. This will give a solution approximately 0.1 (N).

An approximately 0.1 *(N)* solution of sulphuric acid is prepared in exactly the same way as before.

Procedure for standardization of the acid

a) Take a clean burette, pour into it about 5 ml of the acid solution, rinse the whole interior surface of the burette with it and drain out completely. Repeat the process thrice. Fill the burette with the acid solution to a little above the zero mark, (if a funnel is used for filling the burette, remove the funnel immediately after the filling). Open the stop-cock momentarily in order to fill the jet with liquid: examine that no air bubbles are enclosed in the jet, if there are bubbles, run out more liquid until they are removed and the jet is completely filled with the liquid. If necessary, refill burette to bring the level of liquid above the zero mark. Now run out the liquid slowly until the bottom of the meniwng, just touches the zero mark. Now clamp the burette vertically to a burette stand.

   To read the burette, the eye must be at the same level with the meniscus to avoid error due to parallax. Acid is usually taken in the burette.

b) Take a 25 ml clean pipette, dip the lower end into the standard alkali solution, suck in 2-3 ml of the solution and rinse process thrice. Fill the pipette with the alkali solution by suction to a little above the mark, close the upper end with the index finger, wipe out the adhering liquid from the outside of the lower stem. Now relax the pressure of the finger slightly so as to allow the liquid to drain out drop by drop until the bottom of the meniscus just touches the mark in the stem. If any drop is adhering to the tip of the burette remove it by gentle tapping. Now remove the pipette to a clean beaker or a conical flask the tip touching the inner wall of the vessel and run out the liquid into the beaker. When the liquid has run out, hold the jet in contact with the wall for about 15 seconds more and then remove the pipette.

Now place the beaker containing Na_2CO_3 solution on a piece of white paper just below the burette. Add a drop of methyl orange (or methyl red) indicator and run in the acid slowly from the burette, stirring the liquid in the beaker all the while by gently rotating (or shaking) the beaker, until the colour becomes faintly yellow; continue addition of the acid carefully, drop-wise stirring until the colour of end point. Repeat the process two or three times.

[*N. B.* The first titration may be rough one and will indicate the limits (within 0.5 ml) between which the correct end point will lie. The 2nd and 3rd titrations should agree and should not differ by more than 0.1 ml]

$$\text{Strength of the acid soln.} = \frac{\text{Vol. of alkali X Strength of alkali}}{\text{Volume of acid}}$$

The HCl or the H_2SO_4 solution may also be standardized with standard borax solution – the procedure is exactly same as above.

Preparation of standard solution of Caustic soda

Sodium hydroxide is not obtained pure; so a standard solution cannot be prepared by dissolving weighed quantity of the reagent.

Weight out in an ordinary balance about 2 grams of caustic soda take it in a beaker, rinse the piece of solid with a little distilled water, then dissolve it in distilled water. Transfer the solution into a 500 m flask and make up the solution to 500 ml with distilled water. Shake the solution and allow to stand. The solution is approximately 0.1 (N).

Standardization: Pipette out 25 ml of the NaOH soln. into a beaker as before. Add 1 drop of phenolphthalein indicator; the solution turns pink. Titrate with standard HCl or H_2SO_4 solution from the burette with continual stirring; the pink colour becomes fainter and fainter until it is very faint. Continue adding acid drop-wise until with one drop the pink colour just disappears. Note the volume of acid added.

$$\text{Strength of NaOH soln.} = \frac{\text{Volume of acid X Strength of acid}}{\text{Vol. of NaOH soln.}}$$

Solutions of acetic acid & other organic acids are standardized by titration against standard caustic soda solution using phenolphthalein as indicator.

Determination of Sodium Carbonate and Bicarbonate in a mixture

Method I. Procedure : Pipette out 25 ml of the solution in a clean beaker. Add 1-2 drops of phenolphthalein indicator and titrate with standard hydrochloric acid from the burette with constant stirring until the pink colour just disappears. Note the volume of the acid added (V ml). This corresponds to half the carbonate.

$$Na_2CO_3 + HCl \rightarrow NaHCO_3 + NaCl$$

The moment Na_2CO_3 is completely transformed into $NaHCO_3$, the pH of the soln. falls below 8.5 and so phenolphthalein then can not indicate any colour.

Pipette out another 25 ml of the mixed alkali in a clean beaker; add 1 drop of methyl orange and titrate with the standard acid until the colour changes to orange. Note the volume (v ml), this corresponds to total alkali, carbonate + bicarbonate. Let the strength of the acid be S_1 (N).

Calculations

Na_2CO_3 in 25 ml of the mixed alkali = 2V ml of acid S_1 (N)

$=2VXS_1$ ml (N) Na_2CO_3 $=2VXS_1X0.053$ grams of Na_2CO_3 $NaHCO_3$ in 25 ml of mixed alkali

= (v-2V) ml of S_1 (N) acid = (v-2V)XS_1X0.084 grams of $NaHCO_3$

[Eq. wt of Na_2CO_3=53. Eq. wt of $NaHCO_3$=84].

Method II. Procedure : Pipette out 25 ml of the mixed alkali in a clean beaker and titrate in the usual way with standard acid [S_1(N)] using methyl orange indicator. Note the volume (V ml). This corresponds to total alkali, carbonate + bicarbonate.

[VXS_1 ml (N) soln.]

Pipette out another 25 ml of the mixed alkali in a clean beaker. Add to it from a pipette measured volume of standard NaOH soln. [S_2(N)] in excess over that required to transform the bicarbonate to carbonate [$NaHCO_3$+NaOH=Na_2CO_3+H_2O]. Let the vol. of NaOH soln. be V_1 ml. Heat the solution to boiling and to this hot solution add a 10% solution of $BaC1_2$ until the precipitation of $BaCO_3$ is complete. Allow the $BaCO_3$ ppt. to settle. Now titrate the excess of caustic soda immediately without filtering. With a standard soln. [S(N)] of

acid until the pink colour just disappears. Note the volume of the acid (V_2 ml).

Calculations

Let the strength of the acid be S_1 (N)

Let the strength of the acid NaOH solution be S_2 (N)

therefore NH_3 in 25 ml soln. = $(V_1XS_2 - V_2XS_1)$ X 0.0170 gm.

Method III.: 25 ml of the solution is mixed with excess of conc. of NaOH and distilled. NH_3 is quantitatively expelled and is absorbed in a measured quantity (V_1 ml) of standard acid. The excess of the acid is titrated with standard NaOH soln.

Calculations

Let strength of the acid = S_1 (N); vol. of acid = V_1ml

Let strength of NaOH soln. = S_2(N). vol. of acid NaOH soln. = V_2 ml

therefore NH_3 in 25 ml soln. = $(V_1\ S_2 \text{-} V_2S_1)$ X 0.0170 grams.

Oxidation and reduction reactions employed in analysis

Introduction

Processes involving loss of electrons called oxidations, and those involving gains are called reductions. The substance which looses the electrons is called the **reductant** and that which gains them, the **oxidant.** In the course of a reaction, the oxidant is reduced; and the reductant is oxidized. For example:

Fe^{+++} +e → Fe^{++} Sn^{+2}-2_e → Sn^{++++} Ag^+ + e →Ag

$0_{x.}$ Red. Red. Ox. Red. Ox

Electrons do not exist in the Free State; for an oxidant to be able to accept and fix them, a reductant, capable of donating them must be present. For example:

$$2Fe^{+++} + 2e \rightarrow 2Fe^{++}$$

$$\underline{Sn^{++} - 2e \rightarrow Sn^{+++}}$$

$$2Fe^{+++} + Sn^{++} \rightarrow 2Fe^{++} + Sn^{+++}$$

The general equation for oxidation reduction system is Ox.+ne = Red, So, the general equation fro redox potential for a particular system will be, :

$$E = E_0 + \frac{0.06}{N} \log \left[\frac{Ox}{Red}\right] \text{ at } 30°C$$

Where E_0 is a characteristic const. of each system called the normal potential of the system. They are expressed in volts and are referred to the normal hydrogen electrode as the standard reference electrode. *n* is the number of electrons brought into action and [Ox.] and [Red.] represent activities or usually concentrations of oxidant and reductant potential of a mixture of ferric and ferrous iron, for example, is given by :

Fe^{+++}+e- → Fe^{++} (n = 1)

$$E_{Fe} = E_0 + 0.060 \log \left[\frac{Fe^{+++}}{Fe^{++}}\right]$$

Thus, the value of oxidation potential E_{Fe} can be changed by varying the proportion of Fe^{+++} and Fe^{++}

The intensity of the oxidizing or reducing action of a system is determined by its oxidation potential. An oxidant can react with any reductant which is above it in the table. *However, we shall see, that, in quantitative analysis for the reaction to be practically complete, it is necessary for the potential difference between the two systems to be 0.20 to 0.40 volts according to the number of electrons exchanged.

Oxidation-reduction reactions are frequently employed in quantitative analysis. Reagents used in preparing standard solutions for such analysis are :

i) Oxidising agents : Potassium permanganate, Potassium dichromate, Iodine, etc....

ii) Reducing agents : Oxalic acid, Sodium oxalate, Sodium thiosulphate, Mohr's salt, Arsenious Oxide, etc.

Equivalent weight of oxidizing and reducing agents is defined as the weight which reacts with or contains 1.008 gm. of available hydrogen or 8.0 grams of available oxygen or in more generalize way, it is the molecular wt. divided by the number of electrons which one molecule of the substance gains or looses in the reaction.

Analysis involving use of permanganate

Standard solution of Oxalic acid $C_2H_20_4$, $2H_20$

$$\begin{array}{l} COOH \\ | \qquad + 0 = H_2O + \\ COOH \qquad 2CO_2 \end{array}$$

Since 1 mol. of Oxalic acid combines with 16 gms of oxygen, the eq.wt. of it will be half the molecular wt.

Thus, equivalent wt. of. Oxalic acid $[(CO\ OH)_2, 2H_2O] = 63$.

Weight out about 1.5 grams of A.R. Oxalic acid in a 250 ml flask, dissolve in distilled water and make up the volume to the mark. Let the weight be w grams,

$$\textit{Strength}^{*} \text{ of the acid} = \frac{v \times 4}{63} (N)$$

Preparation of $KMnO_4$ solution and standardization

$2KMnO_4 = K_2O$, $2MnO$, 50. *i.e.,* $2MN^{+7} + 10e \rightarrow 2Mn^{+2}$

Eq.wt. of $KMnO_4 = [2KMnO_4/10] = 31.606$

Potassium permanganate cannot be obtained perfectly pure. So, it is not primary standards.

Weigh about 3-3.2 gms of $KMnO_4$ with an ordinary balance. Take in a beaker, add distilled water, cover the beaker with a clock glass heat the solution to boiling, allow it to cool to room temp. Filter the soln. through a funnel containing glass wool. The filtered soln. should be stored in a clean, glass stopped bottle, and kept in the dark except when in use. The above $KMnO_4$ soln. if diluted to 1 litre gives approximately 0.1 (N) soln.

Procedure of standardization

i) Heat water in the wash bottle and allow it to boil.

ii) Rinse the clean burette thrice with 5 ml portions of $KMnO_4$ solution. Fill up the burette with $KMnO_4$ solution in the usual way upto the zero mark—note the upper meniscus. Examine the jet of the burette is completely filled with the solution and no air bubble is left behind.

iii) Prepare a solution of H_2SO_4 (1:2) : Take two test tubes of water in a beaker and add slowly one test tube of conc. H_2SO_4 to it with stirring and cooling, if necessary (express strength in normality upto fourth place of decimal).

iv) Pipette out 25 ml of standard oxalic acid into a 500 ml beaker. Add 15-20 ml of H_2SO_4 (1: 2) and then add boiling water to dilute it to about 230 ml. Now titrate the solution with $KMnO_4$ solution from the burette. At first add $KMnO_4$ solution small quantities at a time with stirring; the pink colour of $KMnO_4$ will at the beginning take some time to discharge. When some $KMnO_4$ soln. has been added, the pink colour will be discharged quickly. Now add $KMnO_4$ soln. more quickly with stirring. Near the end when the rate of disappearance of the pink colour slows down, add $KMnO_4$ soln. drop-wise with stirring until with one drop whole mass of the liquid assumes a pink shade; the pink colour persists for 30 secs., after which the colour may be discharged again.

N.B. Near the end when the rate of disappearance of pink colour slows down, it is more convenient to add 3 drops of diphenyl amine indicator, and then finish the titration by adding $KMnO_4$ solution drop-wise with stirring until a bluish colour just appears; this colour is not discharged on standing rather deepens.

Note the volume of $KMnO_4$ soln. added. Repeat the operation thrice.

Calculations

Let volume of $KMnO_4$ soln. $= V_1$ ml

Strength of oxalic acid $= S_1$ (N)

Strength of $KMnO_4$ soln. $= \dfrac{25 \times S_1}{V}$ (N)

Determination of Oxalic acid in a sample

Pipette out 25 ml of the oxalic acid solution into a beaker and titrate it as above with standard $KMnO_4$ soln. to a pink colour. Note the volume of $KMnO_4$ solution required.

1 ml of (N) $KMnO_4$ solution= 0.0630 grams of oxalic acid.

To estimate the total quantity in grams of Calcium present in the sample (solution) supplied.

Estimation of Calcium : Method

Calcium is precipitated as calcium oxalate by treating a hot hydrochloric acid soln. with ammonium oxalate or oxalic acid neutralizing with ammonium hydroxide (1: 1).

$Ca^{++} + C_2O_4 + H_2O \rightarrow CaC_2O_4H_2O$. This pptn. of Ca^{++} as oxalate is quantitative at pH of 4.

Procedure

Transfer the solution from the container into a 500 ml beaker. Wash container with about 10 ml portions of distilled water 3 or 4 times and transfer the wash liquid into the beaker. Add 1 drop of methyl orange soln. Add NH_4OH, small quantities at a time until the solution smells faintly of NH_3. Then add HCI solution drop-wise with stirring until a faint pink colour appears then add 5 ml of conc. HCI. Heat solution to boiling To this ho solution add a solution of ammonium oxalate, (about 1 gram dissolved in hot water) slowly and with constant stirring. Now add NH_4OH (1: 1) solution until it smells strongly of NH_3. A white ppt. of Ca-oxalate comes down. Allow the mixture to stand for 20 to 30 minutes over a small flame on an asbestos board. The precipitate will settle down leaving a clear liquid at the top. [Add a few more drops of the ammonium oxalate solution down the inclined side of the beaker to see, if the precipitation is complete].

Filter through a quantitative filter paper (Ba-filter paper No. 42). Transfer the ppt. carefully with hot water into the filter paper. Rinse the beaker with hot water 4-5 times, and transfer the liquid into the filter. Now wash the filter paper and the precipitate on the filter paper by blowing a jet of hot water form the wash bottle. [The water in the filter paper should never be allowed to collect upto the rim of the paper in the filter.] The precipitate and the filter paper must be washed free from oxalate and Cl ions. (Test by $CaCl_2$ and $AgNO_3$ respectively). [To test-collect 2-3 ml of filtrate in a test tube, add a few drops of NH_4OH and then 1-2 ml of clear solution of $CaC1_2$+a few drops of HAc—if no ppt. comes down, it is washed free from oxalate].

Place the beaker in which the Ca-oxalate was precipitated below the funnel—its stem touching the inner side of the beaker. Pierce the filter paper with a glass rod and wash down the precipitate with a jet of hot water completely into the beaker. Then add a little hot dil. H_2SO_4 (1 : 8) into the filter paper to dissolve any adhering oxalate ppt. and wash it down again with hot water. Add enough of the hot dilute H_2SO_4 so that the ppt. of Ca-oxalate dissolves completely giving a clear solution.

Now titrate the solution in the usual way by running standard permanganate solution from the burette until the solution just turns pint. Note the volume of the $KMnO_4$ solution.

Calculations

$CaC1_2 + (NH_4)_2\ C_2O_4 = CaC_2O_4, \rightarrow + 2NH_4C1$ ...(1)

$CaC_2O_4 + H_2SO_4 = CaSO_4 + H_2C_2O_4$ (oxalic acid) ... (2)

$\mathbf{2KMnO_4 + 3H_2SO_4 + 5H_2C_2O_4}$

$\mathbf{= K_2SO_4 + 2MnSO_4 + 2MnSO_4\ 10CO_2 + 8H_2O}$ ... (3)

From eqn. (3) it is evident that

1/3 mol. of $KMnO_4$ = 1/2 mol. of oxalic acid

= 1/2 mol. of CaC_2O_4 (from eaqn. 2) ... (4)

Now since equivalent wt. of permanganate is 1/5 molecular wt. From eqn. (4) we have,

1000 ml (N) $KMnO_4$ soln. can oxidize 1/2 molecular wt. Ca-oxalate which contains 20 gm of calcium.

1000 ml (N) $KMnO_4$= 20 gm. of Calcium

1 ml (N) $KMnO_4$ = 0.02 gm. of Calcium.

Standardization of $KMnO_4$ solution with ferrous ammonium sulphate i. e., Mohr's salt

Eq. wt. of Ferrous ammonium sulphate = 392.13

Weigh out about 9.8 grams of the salt (A. R. quality) into a 250 ml flask; add about 150 ml of 1 (N) H_2SO_4 soln. shake well to dissolve the salt; and then make up the volume upto the mark with 1 (N) H_2SO_4 solution. Shake well, allow to stand. Let the weight of the salt be w.

Strength of ferrous ammonium-sulphate soln. $= \dfrac{w \times 4}{392.13}$(N)

when all the dichromate dissolves, add more water and make up the volume upto the mark. The solution is approximately 0.1 (N).

$$\text{Strength of } K_2Cr_2O_7 \text{ soln. } \frac{w \times 2}{49.04}(N)$$

Determination of ferrous iron, ferric iron and and total iron (Fe^{+3}+Fe^{+2}) in a sample : Ferrous iron

Pipette out 65 ml of the iron solution in a 500 ml beaker; add about 20 ml sulphuric acid, phosphoric acid mixture, dilute with distilled water 10 about 250 ml; add 3 drops of diphenyl-amine indicator (indicator should be freshly prepared soln.), and titrate with the standard dichromate solution from the burette, with constant stirring ;the solution becomes green due to formation of green chromium chloride;at the end point there is a sharp change of colour from green to blue. The blue colour is not discharged but deepens on standing. Note the volume of the dichromate solution.

1 ml of (N) $K_2Cr_2O_7$ soln. = 0.05585 gm of Fe.

Total iron (ferrous + ferric) Procedure

Pipette out 25 ml of an iorn solution in a 500 ml beaker; add one test tube full of conc. HCl, boil and reduce with stannous chloride solution; cool, add 10 ml $HgCl_2$ solution and dilute to about 250 ml (see permanganate method). Add about 20 ml of sulphuric acid, phosphoric acid mixture and then 3 drops of diphenylamine indicator and titrat/with standard $K_2Cr_2O_7$ solution until the colour changes from green to a pale blue colour.

[*N. B.* If the ppt. of Hg_2Cl_2 formed on the addition of $HgCl_2$ solution is light and sliky the endpoint is very sharp. If, however, the ppt is too heavy, the endpoint cannot be correctly determined (see titration with $KMnO_4$ soln)].

Note the volume of $k_2Cr_2O_7$ solution; this corresponds of total iron – ferrous + ferric

Ferric iron: As under $KMnO_4$ titration, content of ferric iron is determined by (1) first determining the ferrous iron, then (2) the total (ferrous + ferric) iron and then (3) taking difference.

❑❑❑

Chapter 5

Soils

Soil is the unconsolidated mineral or organic material on the immediate surface of the earth and serves as a natural medium for the growth of land plants. The soil is characterized by varying types when the unconsolidated mineral or organic matter on the surface has been subjected to and shows effects of formation and environmental factors of climate, macro-and micro-organisms acting on parent material over a period of time.

The Soil Profile

The soil profile is comprised of two or more soil layers called horizons, one below the other, each parallel to the surface of the land. Important characteristics of the various horizons are:

- Soil horizons differ in colour, texture, structure, consistence, porosity and soil reaction.
- Soil horizons may be several feet thick or as thin as a fraction of an inch.

Soil differs from the parent material in the morphological, physical, chemical and biological properties. Also, soils differ among themselves in some or all the properties, depending on the differences in the genetic and environmental factors. Thus, some soils are red, some are black; some are deep and some are shallow; some are coarse-textured and some are fine-textured. They serve in varying degree as a reservoir of nutrients and water for crops, provide mechanical anchorage and favourable tilth. The components of soils are mineral material, organic matter, water and air, the proportions of which vary and which together form a system for plant growth; hence, the need to study the soils in perspective. Rocks are the chief sources for the parent materials over which soils are developed. There are three main kinds of rocks: (i) igneous rocks, (ii) sedimentary rocks and (iii) metamorphic rocks.

The rocks vary greatly in chemical composition and accordingly the soil differs in their properties because they are formed from the weathering of rocks. Weathering can be physical or chemical in nature. The agents of physical weathering are temperature, water, wind, plant and animals while chemical processes of weathering are hydration, hydrolysis, carbonation, oxidation and reduction.

A developed soil will have a well-defined profile which is a vertical section of the soil through all its horizons and it extends upto the parent materials. The horizons (layers) in the soil profile which may vary in thickness may be distinguished from morphological characteristics which include colour,texture, structure, etc. Generally, the profile consists of three mineral horizons - A, B and C.

The A horizon may consist of sub-horizons richer in organic matter intricately mixed with mineral matter. Horizon B is below A and shows dominance of clay, iron, aluminium and humus alone or in combination. The C horizon excludes the bedrock from which A and B horizon are presumed to have been formed. A study of the soil profile is important from crop husbandry point of view, since it reveals the surface and the sub-surface characteristic and qualities, namely depth, texture, structure, drainage conditions and soil moisture relationship which directly affect the plant growth. A study of soil profile supplemented by physical, chemical and biological properties of the soil will give full picture of soil fertility and productivity. Physical properties of the soil include water holding capacity, aeration, plasticity, texture, structure, density and colour, etc. Chemical properties refer to the mineralogical composition and the content of the type of mineral, such as Kaolinite, illite and montmorillonite, base saturation, humus and organic matter content. The biological property refer to a content of extent and types of microbes in the soil which include bacteria, fungi, worms and insects.

A. Major Soil Types in India

Some dominant groups of Indian soil, classified according to soil taxonomy and chemical property are mentioned below:

1. Red Soil: They are quite wide in their spread. The red colour is due to diffusion of iron in the pofile.
2. Lateritic soil: are composed of a mixture of hydrated oxides of aluminium and iron with small amounts of manganese oxide.

3. Black soils: This contain a high proportion of Calcium and Magnesium Carbonates and have a high degree of fertility.
4. Alluvial soils: This is the largest and agriculturally most important group of soils.
5. Desert soils: This occur mostly in dry areas and important content is quartz.
6. Forest and Hill Soils: This is high in organic matter.

Taxonomically these soils are classified as follows:

Red soil: Alfisol, Inceptisol and Ultisol

Lateritic soil: Alfisol, Ultisol and Oxisol

Black soil: Vertisol, Inceptisol and Entisol

Alluvial soil: Entisol, Inceptisol and Alfisol

Desert soil: Entisol and Aridisol

Forest and Hill soils: Alfisol

Problem soils: According to salt content:

i) Saline, saline-alkali and Alkali soil

ii) Acid soil

iii) Peaty and Marshy soil (are those soils included in problematic soil a/c to salt content)

The soils are studied and classified according to their use which is termed as land capability classification. In this classification, inherent soil characteristic, external land features and environmental factors are given prominence. For this purpose, soil survey is carried out to record the crop limiting factors, such as soil depth, topography, texture-structure, water holding capacity, drainage features, followed by evaluation of soil fertility status, based on soil testing/analysis. The soils are, thus, classified in to 8 classes, four of which are considered suitable for agricultural purpose (Class I & IV) and Class V to VIII are non-arable lands and can be used for silviculture and forest and need strong conservation measures. An effective linkage between soil testing and soil survey is useful to ensure formulation of a sound soil fertility evaluation programme.

B. Soil Composition

i) Solid phase

- Mineral constituents including weathered rock fragments.
- Organic matter, both dead and living micro and macro organisms.

The mineral constituents are :

Sand - Large particles which are coarse, and individual particles are easily visible (0.02-2 mm in diameter) most common mineral component is silica (SiO_2)

Silt - Medium-sized particles which are 0.002-0.02 mm in diameter.

Clay - Small particles which are less than 0.002 mm in diameter and are referred to as soil colloids.

By definition, particles larger than sand are not soil.

ii) Liquid phase

- Soil water which occupies the pore spaces (between mineral particles) carries the plant nutrients to the roots.

iii) Soil air

- It occupies pore spaces similar to soil water. It fills these voids when soil water is absent.
- It carries respiratory products of roots and soil-organisms.
- It has a higher concentration of carbon dioxide than atmospheric air.

iv) Soil texture

- Refers to the relative proportion of sand, silt and clay present in a soil.
- Based on these proportions soils are classified into various textural classes.
- Clayey soils have a larger per cent of clay. They are considered more fertile than sandy soils but are difficult to work.

- Sandy soils are easy to work but are less fertile. They have low water retention capacity.
- Loamy soils are in between sandy and clayey soils. They are best for arable cropping.

v) Soil structure

- Refers to the arrangement of the different particles into soil aggregates.
- Roots move between these aggregates.
- A compact soil will resist root movement.
- The organic matter content helps the soil aggregation process.

Chapter 6

Soil Fertility

It is the capacity of a soil to supply plant nutrients in adequate amounts to facilitate optimum growth and obtaining the yield potential of a crop. Many soil properties which determine the availability of plant nutrients and thus, soil fertility are as follows:

i) Soil colour

Reddish to brownish colour shows well-drained conditions. Degree of yellowness and mottling show poor drainage. Grey to dark colour indicates the presence of organic material.

ii) Soil depth

Refers to the depth to which plant roots can penetrate the soil and is the distance between the lowest and the upper most horizons of the soil.

iii) Bulk density

It is the weight of the soil in the given volume. A compact soil has a higher value while an organic soil has a lower value. It also affects water holding capacity of the soil.

iv) Field capacity

- Refers to the moisture content of a soil after the loss of gravitational water.
- At this point, water is held in soil micropores, which is available to plant roots, until the water content down to a lower value.
- This lower value is referred to as the permanent wilting point.
- The amount of water available between field capacity and permanent wilting point is referred to as available water.
- This value is important in determining irrigation intervals.

Fate of organic material added to soils

- Undergoes decomposition (only bio-degradable materials) through soil macro and micro fauna and flora.
- Final degradation is made by soil micro-organisms.
- The final product is a humus type material, which has a C:N ratio around 9:1 - 12:1.
- All complex carbon products are converted into simple compounds

x) Nitrogen cycle

This is one of the most important naturally occurring events. It is discussed under:

- Nitrogen fixation.
- Conversion of N in soil.
- Mineralization

a) Nitrogen fixation

- Nitrogen is a very inert gas, which constitutes 78% of the atmospheric air; it has low chemical reactivity.
- For use by plants it has to be converted to ammonium (NH_4^+) or nitrate (NO_3^-)
- Only a few micro-organisms can utilize N gas, which is referred to as N fixation.
- Fixation is effected by microbial action, industrial synthesis and high thermal combustion and lightning.

b) Conversion of N in soil

N occurs in soils in the following form :

- Soluble mineral forms, ammonium, nitrate, nitrous oxide (gas)
- Soluble organic compounds, urea, amino acids.
- Living organisms, plant roots, fungi, bacteria, soil animals.
- Insoluble forms, organic nitrogen, ammonia bonded to clay.

Transformation between the different forms is mediated by soil micro-organisms.

c) Mineralization

- Conversion of N in organic residues and soil organic N into soluble forms through mineralization.
- Carbon sources are degraded sources of energy.
- N in excess of microbial need is liberated.

The following sequence of reactions takes place:

i) **Ammonification:** Complex protein compounds are broken down to ammonium compounds by micro-organisms.

ii) **Nitrification:** Ammonium compounds are oxidized to nitrite and to nitrate by two specific types of soil bacteria, *Nitrosomonas* and *Nitrobactor*.

iii) **Denitrification:** The nitrates are reduced to nitrogen gas under poorly aerated conditions through specific micro-organisms.

❑❑❑

Chapter 7

Methodology for Soil Survey

1. Detailed Soil Survey
2. Reconnaissance Soil Survey

Methodology for Soil Survey

Soil survey is done to characterise the soil resources in the field, group them into various classes, show their distribution on a suitable base map and make interpretations based on their potentials for various uses. Based on the scale, purpose and needs of the user agencies, the surveys can be differentiated broadly into Detailed or Reconnaissance Soil Survey.

Detailed Soil Survey

The purpose of detailed soil survey is to delineate similar areas, which respond or expected to respond similarly to a given level of management. This is achieved by studying all the site (like slope of the land, erosion, drainage, salinity, occurrence of rock fragments, etc.) and soil characteristics (such as depth, texture, colour, structure, coarse fragments, soil reaction, etc.) in detail and grouping/delineating homogenous areas (fields or survey numbers) based on their similarity or difference into various management units.

For Farm level planning, cadastral maps showing all the survey numbers in a village is the ideal base. Cadastral maps are available at different scales (from 1: 4000 to 1: 16000) in our country and they are to be reduced to a uniform scale (preferably to 1:12,500 scale) before the start of the field work, apart from facilitating generation. High-resolution large-scale (1: 12,500) satellite images available from IRS can also be used as a base map but due to the absence of field boundaries; they can be used only in conjunction with cadastral maps for field survey.

Before starting fieldwork in individual villages, preliminary traverse of the block is to be carried out, either by using 1: 50,000 scale toposheet or imagery. During the traverse, based on geology, drainage pattern, surface features, slope characteristics and land-use, landform units and physiographic divisions occurring in the area are to be identified and initial legend is to be prepared by using the available survey information and studying soils at few selected places. This preliminary legend forms the basis to initiate detailed field investigation at the village level later.

For each village, before initiating the field work, the habitations are to be marked clearly by hatched lines, roads by red and all water bodies (rivers/streams/nallas/tanks) by using blue colour pencils. After this, if there is more than one physiographic division in one village, like marine and riverine areas, then they are to be separated accurately on the cadastral map, either by traversing the entire village or by using satellite images. Then within each physiographic division, the various landforms like the hills, hillocks, ridges, dunes, uplands, plains, valleys, etc. are to be identified, their boundaries checked by traversing and corrected, wherever necessary.

After this, starting from a known point, if possible, by taking the help of local village officer, and following permanent features like roads, nallas, tanks, etc. the field boundaries and their corresponding survey numbers are to be identified in each sheet. Wherever new permanent features are seen, they are to be incorporated in the cadastral sheets by using appropriate colours and symbols. Further, the extract extent of the habitations and new settlements has to be marked on the cadastral map, since the original map will not show the real extent or expansion of the village in the recent times. In addition to this the occurrence and extent of other features like rock outcrops, gullies and ravines, quarried areas, fish ponds, etc. has to be marked on the map. The result of this exercise will eliminate the areas that are not to be surveyed and indicate only the areas to be surveyed in each village.

After careful delineation of the various landforms of the cadastral map, updating the cadastral sheets with new permanent features and familiarisation of the area with survey numbers and field boundaries, intensive traversing of each landform (like ridges, uplands, lowlands/ valleys, etc.) is undertaken to select representative areas for transect study. Transects are located across the slope at right angles to the contours and cover most of the variations or changes observed in a landform.

In the selected transect, profiles are located at closely spaced intervals to take care of any change in the land features like break in slope, erosion, gravels, stones, etc. In the selected sites, profiles (vertical cut showing the soil layers from the surface to the rock or water) are to be opened upto 200 cms or rock or hard substratum, whichever comes earlier, and studied in detail for all their morphological and physical characteristics. Detailed soil and site characteristics are to be recorded on a standard proforma for all profile sites as per the guidelines given in the Field Guide. Apart from the transect study, few additional profiles can be studied at random between transects and in the areas not covered by any transects, to bring out all the possible variability of the survey area. Wherever possible, augering can be done during mapping in the field.

If the area is a plain (like in a delta or alluvial plain) or if the slope gradient is negligible, then location of profiles in transects is not possible. In such situations either free style survey or grid type of survey is recommended to locate the profiles. In a free style survey, profiles are located based on the variations observed at the land surface like colour, texture, gravels and stones, erosion status, salinity, wetness, land cover and use, etc. The effectiveness of this method depends on the experience and inference level of the surveyor. In a grid survey, profiles are located at a pre-determined distance irrespective of the variations encountered in the field. Though the number of profiles studied will be more in a grid type of survey it is the easiest one to follow.

Based on the soil-site characteristics recorded from transects study and from random observations, the soils are grouped into different soil series (soil series is the most homogenous unit having similar horizons and soil properties and behaves uniformly for a given level of management). The important characteristics used to group the pedons into different soil series are soil depth, texture, colour, amount and nature of coarse fragments present, calcareousness, presence of calcium carbonates, mottles, coats, stress features, concentrations, nature of the substratum and horizon sequence, etc. A significant difference in any one of these characteristics can be the basis for recognising a different series.

For the soil series identified in the village, the characteristics that differentiate one series from another are listed in the form of a table. In any survey area, the differentiating characteristics between one soil to another will be only few in numbers and this table helps the surveyor to identify and remember them easily. This table is used as a key for grouping any new profile, as and when studied in the area, into the identified series based on their differentiating characteristics.

After grouping the soil into different soil series within each landform, soil mapping is done by intensive traversing on foot covering all plots and survey numbers. During the traverse, starting from one transect area, boundary between two series is established by checking for the differentiating characteristics of one series form the other through a mini pit, auger, road cut or any other information available from the area. This exercise is continued and boundaries established for all the series occurring in the area.

During this exercise itself, within the delineated area of a soil series, based on variations in the surface texture, slope, erosion, presence of gravels, salinity, etc., phases of soil series are separated and their boundaries delineated on the cadastral map (phase is a subdivision of a soil series based on features that affect its use and management. For example, slope, erosion and stoniness). The delineated boundaries are checked on both sides either by traversing the entire course in the field or by using the image characteristics. After completing the mapping of one cadastral sheet, the adjoining sheets are taken up one after the other for detailed mapping.

The separation of phases based on slope, erosion, surface soil texture, gravels, stones, salinity, wetness, etc. can also be attempted at the time of initial traversing of the area itself. If this is done at that stage, then so much time can be saved and only separation of the series boundary is necessary after the study of the profiles and grouping them into various soil series.

A grouped area based on some similarities and marked on the map is called a delineation. A collection of delineations having similar composition is called a mapping unit. The composition of each mapping unit should be described in detail, including any inclusions present, at the time of mapping itself in a field notebook.

The delineated mapping units occurring in a village are shown on the map in the form of symbols. In arriving at the phases, a combination of letters, both in upper (capital) and lower case and numbers are used. For example, the map unit SmcB2, occurring in Kudankulam village, is a phase of Subramaniyapuram series. In this, the first two letters (Sm) indicate the name of the soil series, the third lower case letter c, indicates the texture of the surface soil, the fourth capital letter (B) indicates the slope of the land and the fifth numeral (2) indicates the degree of soil erosion occurring in the map unit. Any other features observed in the field (salinity, gravels, etc.) can be shown by using appropriate symbols on the map.

Collection of other particulars from the survey area

Data on climate, land use pattern, crops grown at present, yield particulars, socio-economic situation, marketing facilities available and other particulars, both at the village and block level, are to be collected during the course of the field survey.

Collection of soil samples

For the soil series identified, soil samples (about 1 kg air dried soil) are collected from a representative pedon for laboratory characterisation from each village. In the surveyed village area, some soil series may occur extensively and in such a situation more than one pedon samples are to be collected for analysis.

Finalisation of block/district level soil maps

From the village maps, soil map of the block or district can be prepared by combining all the village soil maps in a GIS framework. Since the village maps are larger (1:12,500 scale) with phases as mapping units, they are subjected to both cartographic and categorical generalisation to prepare the block level map, preferably at 1:25,000 or 1: 50,000 scale. Due to this generalisation, the block/district level map will have either mono series or association of soil series or even soil complexes as mapping units.

Reconnaissance Soil Survey

The difference between the detailed and reconnaissance survey is the scale of mapping and consequently the accountability of the soil variability encountered in the field. Otherwise, the principles involved and methods used in identifying landforms, transect selection, pedon description, soil series identification, classification and mapping of soils are same in both reconnaissance and detailed surveys. In reconnaissance survey, about 5 per cent of the area is surveyed in detail (like in detailed soil survey) and in the rest of the area, the information is extrapolated.

Sequence of Activities in Detailed Soil Survey

1. Collect all the cadastral maps of the area and bring them to a uniform scale, preferably to 1:12,500 scale, and use it as a base for soil mapping.
2. Take two copies of the base map and mark permanent features like water bodies, streams, etc. with blue colour and roads and habitations with red colour. Other features like rocky areas, forests, salt pans, burial ground, etc. can also be marked.
3. Use one copy of the base map for traversing, updating the map, location of profile sites, delineation of phases and preparation of soil map and another for marking land uses observed in individual fields.
4. The marked copy is to be used for traversing. During the traverse mark the actual extent of habitations and new permanent features like roads, tanks, etc.
5. Start the traverse from a known point. During the traverse observe any change in slope, surface texture, colour, erosion, gravelliness and stoniness, salinity, alkalinity, etc. and mark their boundaries (phases) on the base map.
6. During the traverse itself select and mark representative areas for transects and profile studies in each landform.
7. On the selected transect, open profiles for study and record the observations in the proforma. Compare each profile with already identified/established soil series as given in the initial legend (prepared in the beginning after rapid traverse of the entire block). If the profiles are not matching with the list of series given, keep it separately and propose a new series as and when sufficient number of profiles are studied.
8. Prepare series identification characteristics table based on the groupings done.
9. Start mapping with the help of the series identification table and by checking with mini pits or auger bores, wherever necessary.
10. List the series as well as mapping units occurring in each village and prepare detailed descriptions for both.
11. Collect soil samples from master profiles of each series in each village and sufficient number of water samples from the block for analysis.

12. Hand over the soil map, land use map, series descriptions, map unit description, list of soil samples collected, field notebooks and Soils Handbook to the Project Leader and move to another village.

❑❑❑

Chapter 8

Site Description

The soil continuum seen at the surface of the earth is formed by the influence of at least five soil forming factors like climate, biota, topography, parent material and through the interactions of multitudes of processes operating in nature in course of time. Since these factors and process are never static in any given place and time; soil formation is a dynamic and continuous one. They form and change continually at different rates and along different pathways. As a result they vary spatially in both, the horizontal and vertical dimensions.

To understand the complex patterns of the distribution of soils on the surface, we need to have a thorough knowledge of the role of different soil forming factors and processes that shaped the evolution of landscapes and the distribution of soils on each landscape. To study and map soils, we must understand not only of what they are, but also how they are related to their adjoining counterparts in the field and the sequence there of. For this we need to know the role of rock types, topography, rainfall, vegetation, land-use, etc. in the formation of different soils and their varied patterns observed in an area.

Further, soils are usually hidden from view and they do not have identifiable outer boundaries like other objects. Instead they grade continuously, one into another, in any landform or landscape. Since these gradations are not at random and can be related to the change in the landform features or site characteristics, the spatial variations observed, particularly the horizontal soil variations, can be related to the changes observed in the landform features (like break in slope or rock outcrops, etc.). Deciphering the soil pattern and identification of soil boundary among the soils is possible only when this relationship is established in the field by the surveyor.

So, to identify homogenous areas in the field, which is the objective of soil survey, we need to have not only a detailed description of the soil characteristics (which is represented by a pedon), but also that of and sedimentary rocks (alluvium, sandstone, laterites and other tertiary formations) observed in the State. Refer such state level maps and indicate the type of rocks observed in the survey area in the space provided. Geology maps of the area can be used as a reference and if there is any doubt in the identification of the rock in the field, then a piece of the rock may be collected for confirmation in the base camp.

Parent Material

The loose unconsolidated mineral material formed by the weathering of rocks, from which the soils form, is known as the parent material of the soil. The parent material is designated as C horizon in the soil profile. The parent material can be grouped into those formed in place through the disintegration and decomposition of rocks (Residual) and those that have been transported from the place of their origin by various agents like water, gravity, etc. (Transported). The presence of multiple parent materials should be described in the proforma (colluviums, over residual).

Transported Material

By Wind	Eolian deposit, eolian sands, loess.
By Mass Movement/Gravity	Colluviums, talus, landslide (debris fall/flow deposit, rock fall deposit, soil fall deposit), mine spoil or earthy fill deposits.
By Water	Alluvium, fluviomarine deposit, marine deposit, swamp deposit, beach sand, estuarine deposit, lacustrine deposit.
Residual or in-place or *in-situ* deposits	Residuum (weathered granite/basalt, etc.), saprolite.

Topography of the Surrounding Area

The surrounding land of the profile will normally have complex slopes and the terms used to describe the topography of the surrounding country are indicated below. In contrast to this only simple slopes are used to describe the location of the profile in the pedon description form.

Level	0-1% slope
Nearly Level	1-3% slope
Undulating	3-8% slope
Rolling	8-16% slope
Hilly	16-30% slope
Steep	30-60% slope
Very steep	>60% slope

Landform

Any physical, recognizable feature on the Earth's surface, having a characteristics shape that is produced by natural processes and mappable at common survey scales. Examples of major landforms are given below :

Marine landforms	Beach, beach ridge, dunes, salt pans, swamp, coastal plain, island, gulf, marsh, etc.
Alluvial landforms	Delta, alluvial fan, delta plain, flood plain, plain, etc.
Inland and Hilly landforms	Mountains, mountain slope, mountain valley, hills, high hills, low hills, hill slope, peak, ridge, dyke, tors, inselberg, escarpment, gorge, ravine, vally, etc.

Micro-features (Also known as micro-relief or micro-topography)

Any discrete, natural or artificial surface features, occupying very small areas on the land surface, which cannot be delineated at the scale of mapping are known as microfeatures. These small features, individually, cover less than 100 m^2 area and the height difference will be within few metres from the ground level. For example, small gullies or sand dunes, if they occur in a very small extent in the survey area are described as microfeatures and if the same occupy large areas, then they are delineated and described as a mapping unit.

The other examples of microfeatures are ridge-and-furrow, erosion rills, bar, anthills, channel, depression, hillock, interdune, intermittent stream, minor scarp, mound, hummocks, dune, gilgai, pond, pool, rill, ripple mark, shoreline, tank, contour terracing, levees and minor land slip features. Describe the nature and frequency of occurrence of such microfeatures in the survey area and the relationship of the profile site to such features in the proforma.

Profile Position

In a hilly area the profile position can be indicated as summit, shoulder, backslope, footslope or toeslope, as the case may be. In uplands, the profile position can be indicated as summits, upper, middle and lower part of the upland and lowlands or valleys.

Soil Slope

Soil slope refers specifically to the slope of the land immediately surrounding the profile (i.e. within 100 m of the profile pit) or representative section of the landscape from which the profile is described. Since soil slope is generally in one direction, it is considered as slope. Slope has gradient, complexity, length, form and aspect.

Slope gradient is the inclination of the surface of the soil from the horizontal. It indicates the direction of the surface water flow at the site. The difference in elevation between two points is expressed as a percentage of the distance between those points. For example, if the difference in elevation is 1 metre over a horizontal distance of 100 metres, then the slope gradient is expressed as 1 per cent.

The slope gradient is measured at the profile site by using Abney level and ranging rods. Observations should be taken facing downslope to avoid any errors associated with the clinometers. The Abney level readings, degrees of inclination or declination, can be converted into slope percentages, and slope classes. The equivalence between percentage gradient, degree of slope angle and class of slope used are as follows:

Class of slope	Range of slope%	Abney level reading
A	0-1	0 to 35 min
B	1-3	35 min to 1 degree 44 min
C	3-5	1 degree 44 min to 2 degrees 52 min
D	5-10	2 degrees 52 min to 5 degrees 43 min
E	10-15	5 degrees 43 min to 8 degrees 32 min
F	15-25	8 degrees 32 mins to 14 degrees 03 mins
G	25-33	14 degrees 03 mins to 18 degrees 16 mins
H	33-50	18 degrees 16 mins to 26 degrees 34 mins

Slope Complexity refers to the relative uniformity (smooth linear or curvilinear = simple slopes) or irregularity (complex) of the ground surface. Complex slopes are group of slopes that have definite breaks in several different directions, which results in different slope gradients within the areas delineated. Complex slopes are used to describe the landform of the surrounding country.

Slope length indicates the distance upto which there is no break in the slope. For example, the length of B slope is 100 m, which indicates that the distance between the starting point of the slope and the point where it breaks is about 100 m in length.

Slope form refers to the form of the slope the direction of slope. It is described as convex, concave or straight.

Erosion

The detachment and movement of soil materials from one place to another is known as soil erosion. Depending on the landscape position, vegetation and rainfall, soil erosion may be very slow or very rapid. It is very rapid in the deforested barren hills of the Western Ghats and very slow in the delta or gangetic plains. The agents responsible for erosion are water and wind. Sheet, rill and gully erosions are very widespread in our country.

Sheet erosion is responsible for the almost uniform removal of soil from an area without leaving any significant marks at the surface. It is difficult to notice in the early stages, but it is a serious one on many upland soils.

Rill erosion is the removal of soil through many small incipient channels or rills. It is intermediate between sheet and gully erosion.

Gully erosion is the consequence of water that cuts down into the soil along the line of flow. Gullies form in exposed natural drainage ways, in plough furrows, in animal trails, in roads, between rows of crop plants and below broken man-made terraces. In contrast to rills, they cannot be obliterated by tillage.

Wind erosion is a serious problem in the arid and semi-arid regions, particularly in the rainfed tracts of the country.

Soil erosion classes

The erosion classes are estimated in the field based on the proportion of upper horizons/layers that have been removed (Fig. 2.5). Since these horizons may range widely in their thickness, estimating the absolute amount of erosion in the field is not possible. The erosion classes indicated below are applicable for both water and wind erosion :

Class 1 (Slight erosion) : This class consists of soils that have lost some, but on the average less than 25 per cent, of the original A and / or E horizons or of the uppermost 20 cm of the original A and/or E horizons, if they were less than 20 cm thick. Evidence for class 1 erosion includes (1) a few rills, (2) an accumulation of sediment at the base of slopes or in depressions, (3) scattered small areas where the plough layer contains material from below, and (4) evidence of the formation of widely spaced, deep rills or shallow gullies without consistently measurable reduction in thickness or other change in properties between the rills or gullies.

Class 2 (Moderate erosion) : This class consists of soils that have lost, on the average, 25 to 75 per cent of the original A and/ or E horizons or of the upper most 20 cm, if the original A and/or E horizons were less than 20 cm thick. In cultivated areas with class 2 erosion, the surface layer consists of a mixture of the original A and/or E horizons and material from below. Some areas may have intricate patterns, ranging from uneroded small areas to severely eroded small areas. Where the original A and/or E horizons were thick, little or no mixing of underlying material may have taken place.

Class 3 (Severe erosion): This class consists of soils that have lost, on the average, 75 per cent or more of the original A and/or E horizons or of the uppermost 20 cm, if the original A and/or E horizons were less than 20 cm thick. In most areas of class 3 erosion, material below the original A and/or E horizons is exposed at the surface in cultivated areas; the plough layer consists entirely of this material. Even where the original A/E horizons were very thick, at some mixing with underlying material is common.

Class 4 (Very severe erosion): This class consists of soils that have lost all of the original A and/or E horizons or the uppermost 20 cm, if the original A and/ or E horizons were less than 20 cm thick. In addition, Class 4 includes loss of some or all of the deeper horizons throughout most of the area. The original soil can be identified only in small areas. Some areas may be smooth, but most have an intricate pattern of gullies. Indicate the kind or degree and class of erosion observed at the profile site in the proforma.

Surface Runoff

Surface runoff or external soil drainage refers to the loss of water (includes both rainfall and water flowing from other or nearby soils) from an area by flow over the land surface. Six runoff classes are recognized based on the characteristics of the soil profile, soil slope, climate and vegetative cover :

Ponded : None of the water added to the soil as precipitation or by flow from surrounding areas escapes as runoff. This condition occurs normally in depressions.

Very slow : Surface water flows away very slowly that free water lies on the surface for long periods or enters immediately into the soil. In very slow condition, most of the water either passes through the soil or evaporates into the air. This condition is observed normally in level to nearly level areas or in very porous sandy soils.

Slow : Surface water flows away slowly that free water lies on the surface for significant periods or enters rapidly into the soil. In very slow condition, large part of the water either passes through the soil or evaporates into the sir. This condition is observed normally in nearly level or gently sloping areas or in sandy soils. Normally there is little or no erosion hazard.

Medium : Surface water flows away at such a rate that a moderate proportion of the water enters the soil and free water lies on the surface for only short period. In this condition, large part of the rainfall is absorbed by the soil and for plant growth. The erosion hazard may be slight to moderate when these soils are brought under cultivation.

Rapid : A large part of the rainfall moves rapidly over the surface of the soil and a small part moves through the soil profile. In this condition, water runs off nearly as fast as it is added on the surface. Rapid runoff areas are observed normally in moderately steep areas and in soils with low infiltration capacity. The erosion hazard is normally moderate to high.

Very rapid : A very large part of the rainfall moves rapidly over the surface of the soil and a very small part moves through the soil profile. In this condition, water runs off as fast as it is added on the surface. Rapid runoff areas are observed normally in steep to very steep areas and in soils with low infiltration capacity. The erosion hazard is normally high or very high.

Natural Drainage Classes

Natural drainage class refers to the frequency and duration of wet periods under conditions similar to those under which the soil developed (Table 2.1 and Fig. 2.6). Alteration of the water regime, either through drainage or irrigation, is not a consideration unless the alterations have significantly changed the morphology of the soil.

Ground Water Depth : Indicate the depth of the water table and seasonal fluctuations of the profile site area. The water table measurements can be taken from the nearest open or bore wells or by enquiring with the farmers of the area.

Flooding : Wherever records are available, they can be collected and the frequency can be indicated and in other areas, it can be estimated based on the site characteristics and other converging evidences.

Frequency Class	Criteria
None	No possibility of flooding in the area
Rate	1 to 5 times in 100 years
Occasional	5 to 50 times in 100 years
Frequent	> 50 times in 100 years, i.e. once in two years
Very frequent	Every month > 15 days in a year, used for tidal flooding

Salt / alkali (per cent surface coverage)

The presence of salinity or alkalinity can be identified based on the occurrence of barren areas, presence of salt tolerant crops like prosopis and very poor or stunted growth of plants. Presence of white encrustation on the surface of the soil is an indication of salinity and smooth or fluffy feel to the feet indicates alkalinity in the field. Observe the extent of the area covered by the saline on alkali areas and indicate the per cent surface cover in the column provided.

Soil Reaction (pH)

In the field, pH is estimated by using a variety of field test pH indicator papers and portable pH meter. Depending on the availability and convenience, either of the two methods can be used to find out the reaction of the soil. Though the soil reaction classes used are normally narrow in their range, for field estimation a combination of classes, as indicated in the proforma is sufficient.

Electrical Conductivity (EC)

It is a measure of the concentration of water soluble salts in soils. The occurrence of bare sports, salt tolerant crops like prosopis and uneven crop growth are indicators of salinity and high EC in the field. Portable field EC meters are used to estimate the salt content (extract from saturated soil paste). Wherever field estimation of EC is not possible, try to infer the approximate EC values of the areas and indicate it in the respective column.

Assessment for the surface fragments is done separately for the gravel and for stones and boulders. Indicate the size of the fragments observed in the field. The following classes, based on the areal coverage are used for gravel.

Gravelliness class	% of area covered
Non-gravelly	< 15 per cent
Gravelly	15 to 35 per cent
Very gravelly	35 to 60 per cent
Extremely gravelly	60 to 80 per cent
Considered as part of the top soil	> 80 per cent

The approximate amount of stones and boulders present at the surface has to be assessed separately and reported. The stoniness classes used are indicated below :

Stoniness class	Percentage of surface covered
Stony (class 1)	0.01 to 0.1 per cent of the surface
Very stony (class 2)	0.1 to 3 per cent of the surface
Extremely stony (class 3)	3 to 15 per cent of the surface
Rubbly (class 4)	15 to 50 per cent of the surface
Very rubbly (class 5)	>50 per cent of the surface

Rock outcrops

The distance between the rock outcrops and their percentage coverage in the field is to be recorded as indicated below.

Percent coverage	Description
< 2	No rocks or very few rocks to interfere with tillage
2 to 10	Fairly rocky, sufficient to interfere with tillage but not to make inter-tilled crops impracticable. Exposures are roughly 35 to 100 m apart.
10 to 25	Rocky, sufficient to interfere with tillage of inter-tilled crops impracticable. Exposures are roughly 10 to 35 m apart.
25 to 50	Very rocky, sufficient to make all use of machinery impracticable, except for light machinery. Exposures are roughly 3.5 to 10 m apart.
50 to 90	Extremely rocky, sufficient rock outcrops to make all use of less.
Over 90	Rock outcrops

Elevation and Relief

Elevation refers to the height of a point on the earth's surface, relative to mean sea-level. It can be determined from the topographic maps or by using a calibrated altimeter or by Global Positioning System (GPS). Relief refers to the elevation or differences in elevation above mean sea-level, considered collectively, of a land surface on a broader scale.

Rainfall : Indicate the average annual rainfall of the area in mm

Land Use

Indicate the name of the crop/crops (common names like bajra, ragi, etc. are preferred) cultivated in the current season and wherever possible, the crops cultivated in the previous season, if they are different from the present one. Also, provide the approximate yield/ha and management level (low, medium or high) followed by the farmer for the major crops cultivated. If the crop is irrigated, indicate the method of irrigation and indicate the major and minor crops, if it is a mixed one.

Vegetation

The type of vegetation observed should be described first in simple terms, like evergreen, deciduous or shrub forests, grasslands, etc., and then their common names and their species names, wherever possible.

Generally, a close relationship exists between native vegetation and kinds of sol. The growth and stand of native vegetation of cultivation crops will be of great help in recognizing soil boundaries. Even within a field, differences of vigour, stand, or colour of the crop or of weeds commonly mark soil differences and as such are valuable clues to the location of soil boundaries in the field to the surveyor.

❑❑❑

- Tube sampler: This sampler is useful for rapid sampling of topsoil in medium textured soils. It is not effective on non-coherent soils, hard clays or gravelly soils.
- Blades, shovels, spades, knives, and core samplers.

2. Tubes and cylinders.
3. Hydraulic equipment.

b) Sampling for Soil Fertility Tests

Chemical analysis of soil samples is a prime source of information on the relative availability of plant nutrients. When the soil test is calibrated with crop response and soil samples are properly obtained, the test results can be a strong pivotal point around which management decisions on fertilizer application and soil fertility management can be taken. However, if soil test correlations are weak or soil samples unrepresentative, then the soil test will not be very useful and it may be misleading. This leads to the following questions:

1. Does the soil nutrient level indicated by the chemical analysis actually have a relationship to the nutrient uptake and growth of plants? In other words, does the soil test correlate with fertilizer requirements and crop performance in the field?
2. How well does the soil sample represent the field in question? The analytical aspect of soil testing may correlate perfectly with crop response, but if the soil sample is unrepresentative, then the soil test result cannot have much value.

c) Time of Sampling

The correct time of sampling cannot be decisively specified. However, sampling after fertilization should be avoided. A key point is to collect samples in time to be analyzed for deciding fertilizer application rates and their application. Samples can be taken during crop growing period, when soil content of nutrients is stressed. For example, samples collected in spring, as soil starts warming up, would give lower values of available-P than samples collected in autumn, when soil temperature is low. The reason for the lower values for available-P in spring could be the increased soil microbiological activities, which use up part of the soil-P. It is important to take this into consideration in the evaluation. It is usually recommended that each field be tested once every three years. In most instances, this is often enough to determine whether the fertilization programme is adequate, and whether there is need for an increase or decrease in the rate of application.

d) Depth of Sampling

Samples are usually collected from soil surface layer down to a depth of 15-25 cm in cultivated fields. Some farmers plough to a depth of 30 cm, and then samples should be collected upto this depth. When fertilizers are broadcast on soil surface, still the samples must be collected all through the plough-depth of 15-25 cm. In vegetable fields, it is advisable to collect separate samples from the 0-15 and 15-30 cm layers. In fruit tree orchards, the depth of soil sampling should be extended to the depth of root zone, 0-30, 30-60 and 60-100 cm, if the root system reaches to about 100 cm of depth.

e) Methods of Sampling

A field can be sampled to estimate its fertility through soil analysis by several methods. These include, among others:

1. A composite sample, i.e., a single sample consisting of the composite of several randomly selected soil cores/samples.
2. Multiple samples taken systematically on a grid pattern. The multiple sampling method is the best because, in addition to providing estimates of nutrient levels, it shows variation in fertility from one part of the field to another. Intensive soil sampling will not be practical unless the economic returns justify the costs, or where the topographic conditions are such that high variability is expected. Thus, the most commonly recommended sampling procedure is the composite sample. A good composite sample is made up of a number of soil cores taken at random from 15 - 20 sites in a 1 - 5 ha cultivated field. The purpose of this procedure is to minimize the influence of any local variability in the soil. For example, band-applied fertilizers are not uniformly applied over the soil surface. In addition, there may be some spots in the field where fertilizers were spilled or plant refuse was buried. A sample taken from such an area would be completely misleading. Consequently, most recommendations call for taking borings at 15-20 locations over the field for each composite sample. Areas that vary in appearance, slope, drainage, soil type, and previous

treatments should be sampled separately, and areas that cannot be treated or fertilized separately should be omitted. Shows a sampling pattern in a heterogeneous land. Many growers are directing attention to these small spots in their fields and treating them as needed. Hence, it is advisable to take separate samples from these localized areas of poor crop growth. Sampling recently fertilized bands, dead furrows, areas adjacent to roads, or where manure, compost or crop residues have been piled should be avoided. A random distribution of individual sample sites would be ideal. However, equally reliable information is obtainable with much less effort by reducing the sampling area when the fields are relatively uniform. Sampling areas are, often traversed in a zigzag pattern to provide a uniform distribution of sampling sites. Some of these methods are presented in Figure 2. In fruit tree orchards, particularly those irrigated by a drip system, soil samples should be collected from the root zone of trees, i.e., the zone moistened by irrigation, reaching to the depth of the root system. Trees, around which soil is to be sampled, are selected in a random or systematic pattern. A reasonable number of sites (trees) would range from 5-20% of total number of trees.

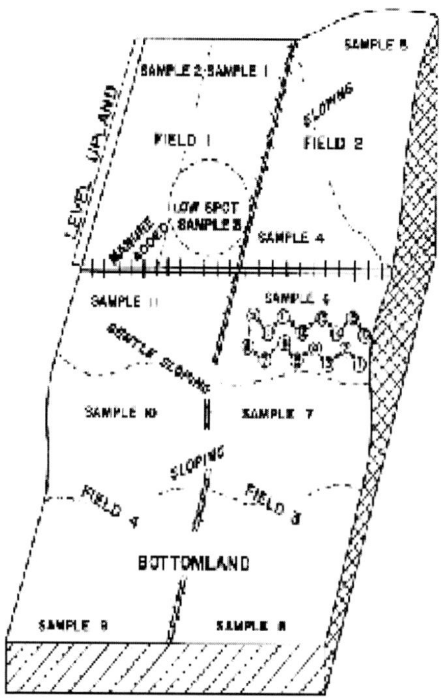

Figure 1: Sampling pattern for fertility test in a non-uniform land

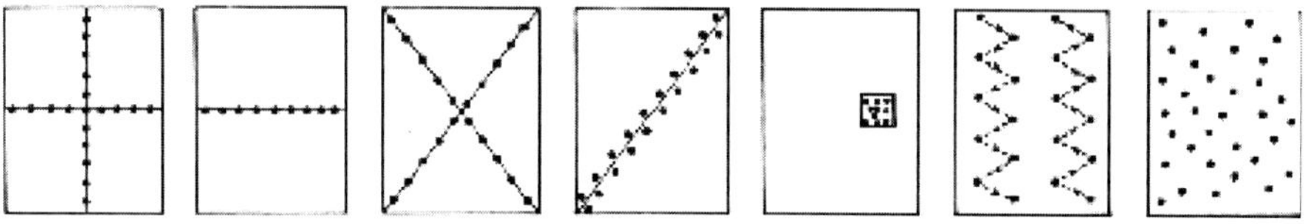

Figure 2: Sampling-sites selection methods for fertility test (Sample numbers refer to composite samples) (*Source*: Tarzi, 1984)

f) Soil Sampling for Soil Survey and Classification Studies

Soil survey is a type of pedological study the purpose of which is to describe soil characteristics for areas of a few to hundreds or thousands of hectares, and delineate these areas on a map. Normally, these surveys are government projects carried out to determine and develop land-use and management. In soils having distinguishable genetic horizons, pits are dug to about two metres depth or to bed rock, if it is shallower than 2 m. Soil profiles are studied and samples are taken from each horizon. Where surface soil did not develop from the underlying parent materials, as in depositional soils, samples are collected from a sequence of depth intervals, e.g. 20-30 cm intervals. Soil surveyors commence from a convenient point or landmark, and then select fields or sites for boring or digging of pits. The methodology will vary in details depending upon the type of terrain, geology, landforms and spatial variability of the land. More important are the survey objectives, the mapping scale, and the complexity of the soil pattern to be surveyed. Arial photos taken for the terrain to be surveyed, and now images taken by remote sensing from satellites, are first studied to collect some basic information about relief and other physical feature of the terrain to be surveyed. The dependability of these images serves in approaching and executing soil survey projects. In the areas with a relatively simple soil pattern, the average observation density will be one pit per 15-20 hectares. In the areas with complex soil patterns, the density will be one per 10-15 hectares. The observation density will be lower in areas of poor soils, such as shallow soils, rocky soils, undulating and ridges. Soil surveyors correlate soil series and mapping units with landform units and land surface features.

g) Preparation of Soil Samples

Following the sampling process, soil cores are placed in a clean container and clods are crushed. After mixing well the gross soil sample, about one kg is transferred into a clean plastic or canvas lined with plastic bag. This is the composite sample to be sent for analysis. Bags should be properly labelled. Information, such as sample number, depth, and date of sampling should be written on the bag from outside, and on a sample card placed inside the bag. Bags should be closed tightly, placed in wooden or cardboard box, and then shipped or taken to the laboratory.

It is essential that a Field Data Sheet be filled and sent with the samples to the laboratory. Air-drying is the most accepted procedure of sample preservation and may reduce the rate of possible reactions in the disturbed soil sample. A soil sample should not be allowed to stay moist for extended periods of time. Soil aggregates should be broken carefully to accelerate the drying process. It is generally assumed that chemical and biochemical reactions in airdry soils are reduced to a minimum, although these reactions are still possible sources of error.

FIELD DATA SHEET

Sample No. ___________

Collector _____________ Address____________________ Date __________

Area _____________ Location_____________ Owner _________________

Farm Size __________________ Vegetative Cover______________________

Source of Water __________________ Water Quality ____________________

Sample Depth ___________________ Previous Crop ___________________

Site Selection Method: Random　　Sample Type:

Individual

Zig Zag　　Composite

Two Way Diagonal　　No. of cores

Two Way Cross-Strip

Test Plot

Purpose of Analysis:

Capability Assessment　　Slope: 1-2%

Fertility Evaluation　　2-5%

Salinity Appraisal　　5-10%

Soil Classification　　10-25%

> 25%

Irrigation Method:

Flood　　Years of Irrigation: Never Irrigated

Furrow　　1-5

Sprinkler　　5-15

Centre Pivot　　> 15

Drip

Rainfed

Years of Cultivation: Never Cultivated　　Drainage: Good

1-5 year's　　Moderate

5-15 year's　　Poor

>15 years

Manure/Fertilizer Application History:

Manure (type) ______________________ Rate (kg/ha) _____________________

Fertilizers (mineral) __________________ Rate (kg/ha) _____________________

Note: Questions for which accurate answers cannot be provided should be left unanswered.

Figure 3: Field data sheet

For some quick analysis live NO_3-N and NH_4-N the measurements are done on moist soil samples as collect farm field. Separate sample is taken to adding the moisture, which is then used to correct results to oven dry basis.

h) Grinding and Sieving of Soil Samples

Thorough mixing requires that the sample be crushed and ground to particles of uniform size. After drying the sample, clods and large aggregates are crushed and mixed. Then the crushed material is further ground to pass 2 mm sieve. Care should be taken not to break the individual soil particles during the grinding process. The purpose of grinding is to reduce heterogeneity and to provide maximum surface area for physical and chemical reactions. Various devices are used for crushing and grinding soils. The choice of equipment depends on four factors:

1. The amount of sample to be crushed or ground.
2. The degree of fineness to be attained.
3. The contamination that can be tolerated.
4. The analysis in question.

Soil aggregates can be crushed by using jaw crushers, hardened steel mortars or rocking boards (Tan, 1996). Jaw crushers are used for crushing large aggregates, steel mortars for smaller aggregates, and rocking boards for small aggregates. The types of grinders used are plate grinders (friction mills), ball mills, rod mills, agate mortars, and boron carbide mortars. For large amounts of samples, plate grinders or ball mills or rod mills are used. The sample is mixed with stainless steel rods or balls, in a container and the container is rolled over a period of time. When properly operated, rod and ball mills are efficient grinders). For least contamination and for small amounts of samples, an agate or boron carbide mortar can be used.

After grinding, the soil is sieved through a 2 mm sieve and stored in airtight glass or clean plastic containers as soon as possible to avoid adsorption of gases in the laboratory. This will serve as stock samples for analysis. Then, samples should be stored in a cool and well-ventilated room. Analysis of samples that are passed through sieves of unspecified sizes may produce results that would not be comparable to internationally accepted data. Therefore, the sieve size should be reported with the results of analysis.

i) Reduction of Sample Size

When a very large sample is collected, it is necessary to reduce the size of the sample for ease of storage and handling. The bulk sample must be reduced in size, and the sub-sample taken for analysis should represent the soil characteristic of the field under investigation. In order to achieve this goal, a random and unbiased method of sub-sampling is essential. Sample splitting can be performed with a mechanical sample splitter, This process can be repeated as many times as necessary. Another way for reduction of sample size is by *quartering*. The sample is spread uniformly over a plastic sheet and divided into four equal portions. Portions 2 and 3 are collected and thoroughly mixed, whereas the remainder is discarded. This process of quartering can be repeated as many times as necessary until the proper size of sample is attained.

❑❑❑

Chapter 10

Physical Properties

1. Soil Moisture

Water is held in soil by adhesive and cohesive forces. Adhesion is the force of attraction between the solid soil particles and water molecules. Cohesion is the mutual force of attraction between water molecules. Another force that affects water retention and movement in soil is the capillary force, which is a function of the size of micropores or capillaries.

The pF is the term mostly used to express soil moisture, although other terms are also used, like atmosphere or bar. Water can exist in soil under a tension varying from pF = 0 (no tension) to pF = 7.0 (high tension). The point at which water is held in soil after excess water has drained out by gravity is called field capacity (pF = 2.54). The point at which water is held in soil at a force that plants cannot extract at a sufficient rate to maintain turgor, is known as wilting point (pF = 4.2). At this point the plants start to wilt. The amount of water between field capacity and wilting point is called available water. Water in air dry soil is held with a tension of pF 6.0.

Apparatus

- Moisture determination cans
- Analytical balance
- Drying oven
- Desiccator

Procedure

1. Weigh accurately a metal can with lid (W1).
2. Place about 50 g of soil in the can and weigh accurately along with the lid (W2).

3. Place the can with the lid under it in a drying oven at 105°C for 24-48 hours, or until constant weight is reached.
4. Remove the can from the oven, cover it tightly with the lid, and place in a desiccator to cool.
5. After cooling, weigh the can accurately with the oven-dry soil in it. Record the weight (W_3).
6. Compute per cent moisture content on oven-dry basis.

Calculations

Weight of water = W_2-W_3

Weight of oven dry soil = W_3-W_1

$$\text{Moisture \% (oven - dry basis)} = \frac{W2-W3}{W3-W1} X\ 100$$

The corresponding correction factor or multiplication factor for amount of sample to be weighted in for analysis is :

$$\text{Moisture correction factor} = \frac{100 + \%\ Moisture}{100}$$

Weight of water = W_2-W_3

Weight of wet soil = W_2-W_1

$$\text{Moisture \% wet-soil basis} = \frac{W2-W3}{W2-W1} X\ 100$$

Gravimetry with drying by burning

Alcohol or spirit (Bouyoucos, 1937)

In this method, the soil moisture is evaporated by igniting the soil mass with alcohol or spirit. The process is repeated to get constant weight of ignited soil. **Equipments and reagents :**

1. Moisture cans
2. Glass rod
3. Ethyl, methyl or propyl alcohol
4. Match box
5. Balance with weights
6. Measuring cylinder
7. Desicator.

Procedure

Weigh about 20-25 gm of wet soil in a moisture can add about 5 ml alcohol or spirit drop to fully saturate the soil and ignite. Cool the cans in a desicator and record the weight. Repeat the process several times till a constant weight of the ignited soil is recorded. Compute the percentage moisture on the basis of the constant dry weight of the burnt soil.

This method is rapid, reproducible and more suited for field soil moisture determinations where laboratory facilities are not adequate.Burning of the soil results in loss of organic matter.

This method needs to be calibrated against oven dry method for each soil type.

In direct method (By use of neutron moisture probe)

Rather than taking a sample of soil, it is often desired to measure the soil moisture status *in situ* without disturbing the system. The neutron moisture meter is such a device which is much used in the field for measuring water content (Belcher, 1952, Gardner and Kirkhana, 1952, Holmes, 1956 and Van Bavel *et al.*, 1956). The neutron method is a measure of the number of hydrogen nuclei that are present per unit volume of soil. Therefore, it is a means of directly observing the moisture content by volume.

Principle

Hydrogen nuclei have a marked property for scattering and slowing neutrons. This property is exploited in the neutron method for measuring soil water content. High energy neutrons (0.1 to 10 MeV) emitted from a radioactive substance, such as radium beryllium (5Mc) or americium-beryllium (30Mc) neutron source are called fast neutrons and travel with high speed of the order of 100 miles per second.

When such a fast neutron source is placed in a moist soils, the emitted neutrons interact with the surrounding medium. They collide with the nuclei of the soil in a billiard ball fashion, changing direction as a result and loosing energy. With the energy loses, the speed diminishes until it approaches the speed characteristics for molecules at the prevailing temperature. Such neutrons are called thermal or slow neutrons. The slow neutrons are finally absorbed by other nuclei

checks, should be made to each soil type as guided by experience in area. It is very important to realize that the potential precision of the neutron method is greater than that of any other method under field condition. Therefore, calibration should be carried out with large homogeneous masses of soil of which the bulk density and moisture content (by gravimetric method) are accurately known.

2. Grain Size Analysis (Sieve Analysis)

This test is performed to determine the percentage of different grain sizes contained within a soil. The mechanical or sieve analysis is performed to determine the distribution of the coarser, larger-sized particles, and the hydrometer method is used to determine the distribution of the finer particles.

Grain size analysis provides the grain size distribution, and it is required in classifying the soil.

Equipment

Balance, Set of sieves, Cleaning brush, Sieve shaker, Mixer (blender),

Proocedure

Sieve Analysis

1. Write down the weight of each sieve as well as the bottom pan to be used in the analysis.
2. Record the weight of the given dry soil sample.
3. Make sure that all the sieves are clean, and assemble them in the ascending order of sieve numbers (#4 sieve at top and #200 sieve at bottom). Place the pan below #200 sieve. Carefully pour the soil sample into the top sieve and place the cap over it.
4. Place the sieve stack in the mechanical shaker and shake for 10 minutes.
5. Remove the stack from the shaker and carefully weigh and record the weight of each sieve with its retained soil. In addition, remember to weigh and record the weight of the bottom pan with its retained fine soil.

Data Analysis

Sieve Analysis

1. Obtain the mass of soil retained on each sieve by subtracting the weight of the empty sieve from the mass of the sieve + retained soil, and record this mass as the weight retained on the data sheet. The sum of these retained masses should be approximately equals the initial mass of the soil sample. A loss of more than two percent is unsatisfactory.
2. Calculate the per cent retained on each sieve by dividing the weight retained on each sieve by the original sample mass.
3. Calculate the per cent passing (or per cent finer) by starting with 100 per cent and subtracting the per cent retained on each sieve as a cumulative procedure.

   For example: Total mass = 500 g

   Mass retained on No. 4 sieve = 9.7 g

   For the No.4 sieve:

   Quantity passing = Total mass - Mass retained

   = 500 - 9.7 = 490.3 g

   The per cent retained is calculated as;

   % retained= Mass retained/Total mass= (9.7/500) X 100 = 1.9%

   From this, the % passing = 100 - 1.9 = 98.1 %

   For the No. 10 sieve:

   Quantity passing = Mass arriving - Mass retained = 490.3 - 39.5 = 450.8 g

   % Retained = (39.5/500) X 100 = 7.9 %

   % Passing = 100 - 1.9 - 7.9 = 90.2 %

   (Alternatively, use % passing = % Arriving - % Retained

   For No. 10 sieve = 98.1 - 7.9 = 90.2 %)
4. Make a semi-logarithmic plot of grain size vs. per cent finer.
5. Compute Cc and Cu for the soil.

3. Soil Texture – Particle Size Distribution

a) Introduction

Soil texture, or particle size distribution, is a rather stable soil characteristic which influences the physical and chemical properties of the soil. There is a direct relation between particle size and the total surface area of particles in a given weight of soil. As particle size decreases, total surface area, better known as specific surface area, increases. Since most of the physical and chemical properties of soil are related to surface activity, determination of particle size distribution is a standard procedure for characterizing and classifying soils.

As soil particles are freely settling down the water column, they are sorted according to particle size. Coarse particles, like sand, move down faster than fine particles. Because of their differential settling rate in water, the size fractions are sorted into different size ranges. The Bouyoucos Hydrometer is calibrated to read the amount of solid particles remaining in suspension, hence the method is known as Bouyoucos method. Two basic conditions are required in this method:

- Dispersion of the soil sample into a state of single grain particles
- During sedimentation, particles should be moving down freely

Determination of the particle size distribution of gypsiferous-calcareous soils is a tedious and time-consuming process. Gypsum, which inhibits soil dispersion, is usually removed from soil samples before analysis, although gypsum has marked effects on the physical properties of soils and it is most desirable to determine the particle size distribution without removing the gypsum fraction. So far, no method for the determination of the particle size distribution of gypsiferous-calcareous soils is entirely satisfactory. The estimation of texture under field conditions is misleading because of the presence of gypsum crystals in various sand-sized fractions. The forms and degree of crystallization of gypsum particles influence the feel of the soil and as a result field estimates of texture are generally coarser than indicated by laboratory determinations. Based on the experience gained from the analysis of a large number of gypsiferous soil samples, the following can be Concluded:

1. The sum of the various size-fractions of gypsiferous horizons may be less than 100 per cent due to dehydration of $CaSO_4.2H_2O$ during oven drying at 105°C.

2. The distribution of gypsum within the various size-fractions depends on the total amount of gypsum The difference between the weight of such fractions obtained after treatment at 50°C and 105°C shows the following:
    - The fraction less than 20 μm in diameter is greater, when the total content of gypsum is about 10 per cent, and decreases with the increase of gypsum content to about 25 per cent. This fraction does not change as gypsum content increases from 25 to 45 per cent, but it increases again as gypsum content goes above 45 per cent.
    - The amount of the fractions greater than 20 μm in diameter increases as the total amount of gypsum increases. For most samples with gypsum content between 10 and 25 per cent, the relationship is not significant.
3. Gypsum is found in all size fractions, but it is mostly present in the coarse and fine sand fractions (0.02-2.0 mm) followed by the silt fraction.

b) Dispersion

Individual soil particles must be separated from each other, and kept separated during the determination in order to correctly measure particle size distribution. Since aggregates of solid particles are usually held together by some kind of binding agent, it is first necessary to remove these substances, or at least render them ineffective. Once the soil aggregates are separated into individual particles, they are described as dispersed particles.

Dispersion is achieved by chemical and mechanical means. Sodium hexameta phosphate is an effective chemical dispersing agent for two reasons:

1. The sodium monovalent cation replaces the usual polyvalent cations adsorbed on soils, thereby breaking interparticle linkages. The polyvalent cations are also reduced in activity as a result of reaction with the phosphorus and consequent precipitation.
2. The adsorbed sodium cations raise the electronegativity of colloids until these particles repel each other and remain dispersed. After the binding agents are removed, aggregates breakdown is further facilitated by mechanical stirring. The mixture of dispersed soil particles in water is called a soil suspension.

c) Sedimentation

In sedimentation techniques, the settling rates of dispersed particles in water are measured. Large particles are known to settle out of suspension more rapidly than do small particles. This is because larger particles have less specific surface area and hence less buoyancy than small particles. Stocke's law is used to express this relationship:

$$V = \frac{2}{9} X \frac{r2g\,(d1-d2)}{n}$$

The formula shows that settling velocity, **V**, is directly proportional to the square of the particle's effective radius, **r**; the acceleration of gravity, **g**; and the difference between the densities of the particle, **d1**, and that of the liquid, **d2**; but inversely proportional to the viscosity of the liquid, **n**. Density of water and its viscosity are both affected by temperature, and thus, particles settle faster with increased temperature. This is why it is necessary to apply correction fort temperature.

In practice, the amount of material remaining in suspension at any time is measured with a hydrometer, which indicates the density of the suspension at the hydrometer's centre of buoyancy. Bouyoucos determined that after 40 seconds all sand-sized particles (0.02 mm and larger) settle out of the suspension and do not influence the hydrometer reading. After two hours, particles larger than clay (0.002 mm) settle out of the suspension and do not affect the hydrometer reading any more.

d) Bouyoucos Method

Apparatus

- Electric mixer (stirrer), with baffled stirring cup
- Settling cylinder
- Graduated cylinder, 1 litre, with 1 000 ml mark 36 r^2 cm from bottom
- Bouyoucos Hydrometer calibrated at 200°C
- Thermometer C°, Stop watch

Reagent

- Sodium hexametaphosphate, $Na_6O_{18}P_6$ solution (examples of commercial names: Calgon, Graham's salt, glassy sodium): Dissolve 50 g of Calgon in water and dilute the solution to a volume of 1 litre.

Procedure

1. Weigh 50 g oven dried fine textured soil (100 g of coarse textured soil) into a baffled stirring cup or conical flask. Treat the soil with H_2O_2 to destroy organic matter. Fill the cup to its half with distilled water and add 10 ml of sodium hexametaphosphate solution.
2. Place the cup on stirrer and stir until soil aggregates are broken down. This usually requires 3-4 minutes for coarse textured soils and 7-8 minutes for fine textured clay.
3. Transfer quantitatively the suspension to the settling cylinder by washing the cup with distilled water. Fill the cylinder to the lower mark with distilled water after placing the hydrometer in the liquid. If 100 g of coarse textured sample was used, fill to the upper mark on the settling cylinder.
4. Remove hydrometer and shake the suspension vigorously in a back and forth manner. Avoid creating circular currents in the liquid, as they will influence the settling rate.
5. Place the cylinder on a table and record the time. After 20 seconds, carefully insert the hydrometer and read the hydrometer at the end of 40 seconds.
6. Repeat step 4 and 5 to obtain hydrometer readings within 0.5 g differences from each other. The hydrometer is calibrated to read grams of soil material in suspension.
7. Record the readings on the Data Sheet for Hydrometer Readings (a sample is shown in Figure 5).
8. Measure the temperature of the suspension. For each degree above 20°C add 0.36 to the hydrometer reading, and for each degree below 20°C, subtract 0.36 from the hydrometer reading. This is the corrected hydrometer reading.
9. Re-shake the suspension and place the cylinder on a table, where it will not be disturbed. Take a hydrometer reading exactly two hours later. Correct for temperature as described above.

10. From the percentage of sand, silt and clay as calculated on the Data Sheet, use the diagram for textural triangle (Figure 5) to determine the textural class of the soil :

1. Soil sample identification number ....
2. Soil weight (g) ....
3. Forty second hydrometer reading (g) ....
4. Temperature of suspension (C°) ....
5. Corrected 40-second hydrometer reading (g) ....
6. Two hours hydrometer reading (g) ....
7. Temperature of suspension (C°) ....
8. Corrected 2-hour hydrometer reading (g) ....
9. Grams of sand (the sand settles to the bottom of the cylinder within 40 seconds, therefore, the 40-second corrected hydrometer reading actually gives the grams of silt and clay in suspension. The weight of sand in the sample is obtained by subtracting line 5 from line 2) ....
10. Grams of clay (the corrected hydrometer reading at the end of two hours represents grams of clay in the suspension since all sand and silt has already settled by this time) ....
11. Per cent sand (line 9/line 2) X 100 ....
12. Per cent clay (line 10/line 2) X 100 ....
13. Per cent silt (find the silt by difference. Subtract the sum of the per cent sand and clay from 100) ....
14. Soil class ....

Figure 4: Data sheet for recording of hydrometer readings

Some experienced field men take a sample of soil in their hand, feel it with their fingers when dry, and then inspect its stickiness after wetting, and then they come up with a skilled guess about the textural class of the soil.

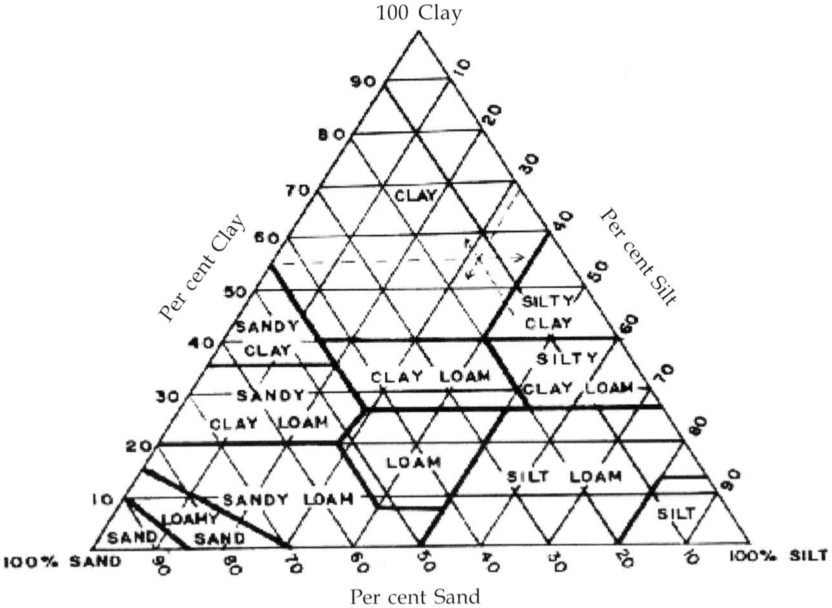

Figure 5: Texture triangle showing the percentages of sand, silt and clay in the textural classes (the intersection of the dotted lines shows that a soil with 55 per cent clay, 32 per cent silt and 13 per cent sand has a clay texture).

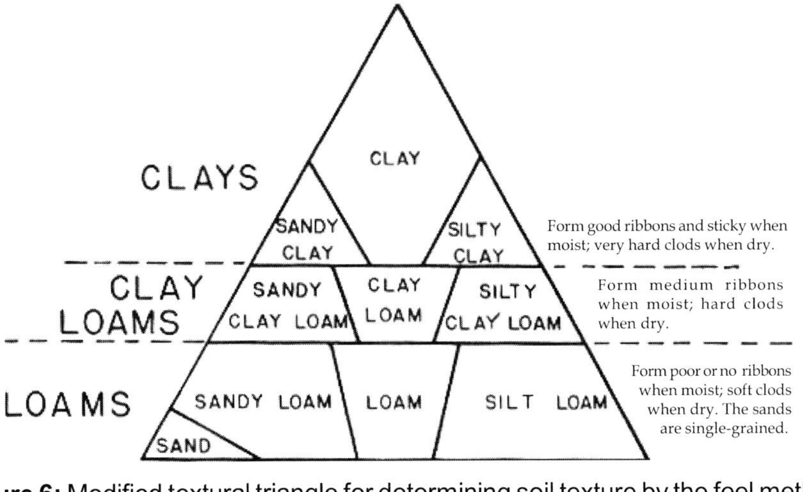

Figure 6: Modified textural triangle for determining soil texture by the feel method (*Source:* Foth *et al.*, 1977).

3. Soil Density and Total Pore Space

6.1. Introduction

Soil bulk density (BD) is defined as the mass of oven-dried solids divided by the bulk volume of the solids plus pore space at a specified soil water content, usually the moisture content at sampling.

Soil particle density (PD) is defined as the mass of oven-dried solids divided by the volume of the solid particles alone. Actually, it is the density of the soil particles, irrespective of the volume of voids between particles. The magnitude of PD depends on the type of minerals in the particles, and the content of organic matter in the soil. The particle density of most mineral soils is about 2.65 g/cm^3. It varies little from soil to soil and, therefore, is of little practical significance. It is used with bulk density to calculate soil porosity.

Bulk density is an extremely useful parameter, as it indicates soil structure and void space. It is required to calculate porosity when particle density is known, to convert weights to volumes, and to estimate weights of soil volumes too large to weigh. It is also required to convert weight-based determinations to a volume based figure, which are often, more interesting. For example, the volumetric content of water in a soil layer is obtained by multiplying the gravimetric water content by the product of the bulk density and the volume of the layer. A medium textured mineral soil that is in good structural condition for plant growth contains about 50% total pore space on volume basis. This pore space is important for gas exchange (O_2 & CO_2) between the soil and the atmosphere, and water storage and movement. The total pore space consists of the pore spaces between adjacent sand, silt and clay particles and those between aggregates. Therefore, texture and structure are the main factors governing the amount of pore space in soil. Organic matter affects pore space indirectly by improving structure. Bulk density is an indirect measure of the total pore space in the soil and is also affected primarily by texture and structure. If the aggregation of a particular soil is increased, the total pore space will be increased, and the weight per unit volume or bulk density of the soil will decrease. The bulk density of fine texture mineral soils ranges from about 1.0 to 1.3 g/cm^3, and that of sandy soils ranges between 1.3 and 1.7 g/cm^3. The bulk density of organic soils is usually much less than that of mineral soils and may be as low as 0.4 g/cm^3. Bulk density and total pore space are readily altered by tillage operations.

b) Measurement of Bulk Density

Two methods are commonly used to determine soil bulk density: one uses samples of disturbed soil, and the other uses samples of undisturbed soil. The second method uses consolidated soil masses, like clods and cores.

c) Bulk Density of a Disturbed Soil Sample

This method is used when it is not possible to take a consolidated sample of soil, as in sandy soils and soils of greenhouses and nurseries because they are loose and very friable.

Procedure

1. Fill a pre-weighed 100 ml graduated cylinder with air-dry soil.
2. Compact the soil in the cylinder by tapping the cylinder firmly 15 times on the palm of the hand.
3. Record the volume of the packed soil in the cylinder.
4. Record the weight of the cylinder and the soil.
5. Calculate the weight of the soil.
6. On a separate sample, determine the moisture content of the soil sample and calculate the oven-dry weight of the soil in the cylinder.
7. Calculate bulk density.

Calculations

1. $\text{BD (g/cm}^3) = \frac{\text{weight of oven dry soil (g)}}{\text{Volume of soil Sample (cm3)}}$

Bulk density of general soil is 1.02 to 1.8 g cm^{-3}

Note: Bulk density is commonly calculated on oven-dry basis, but for certain uses it is calculated on a wet-soil basis.

d) Bulk Density of an Undisturbed Soil Sample

Two methods are used to determine the bulk density of an undisturbed soil sample: the soil clod or ped method, and the core method.

The Clod or Ped Method

Reagent

- A resin, such as Dow Saran F 310 (general purpose non-crystalline coplolymer), is dissolved in acetone at a saran: solvent ratio of 1:7. Acetone is flammable, therefore, this mixture should be prepared in a fume hood and it should be kept in a tightly closed container to prevent volatilization.
- Melted wax, by heating on medium flame, could be used as a replacement to the dissolved saran resin.

Procedure

1. Carefully collect soil clods from a soil profile. If roots are present, cut them carefully with scissors.
2. Tie and hold the clod with a fine copper wire and weigh it.
3. Dip the clod in the saran solution, hang it to dry for 30 minutes. Additional saran coatings may be applied to make the clod more water proof.
4. Weigh the coated clod and wire in air (Wa).
5. To determine the volume of the clod, weigh it while suspended in water with a balance that can accept the clod hanging on the balance beam by the thin copper wire (Ww). The drop in weight (Wa-Ww) is equal to the weight or volume of the water displaced, which is the volume of the clod.
6. If such a balance is not available, immerse the clod completely in a graduated cylinder half full with water, and detect the change in volume of water in the cylinder (V).
7. Break the clod and take a sample for the determination of soil moisture content. This is needed to calculate the oven-dry weight of the soil (Wod).

Calculation

$$BD = \frac{Wod}{Wa - Ww}$$

W_{od} is the mass of solid particles in the clod in grams.

$(W_a - W_w)$ is the weight of the water displaced by the clod in grams, or volume of the clod in cm3. If the volume of the clod is measured as indicated in step 6:

$BD = W_{od} / V$

V is the change in volume of water in the graduated cylinder in cm3.

Note: The clod method may yield higher BD values than other methods.

The Core Method

Procedure

1. Carefully drive a thin-walled steel tubing or pipe (of known weight and volume) into the soil with a block of wood and a hammer. Be careful to avoid compaction of the soil during collection of the cores. After careful removal of the soil core, examine it and trim the ends carefully.
2. Weigh the soil and tubing. Calculate the weight of the soil sample alone by subtracting the eight of the steel tubing.
3. Take a portion of this soil for the determination of soil moisture.
4. Knowing the moisture content of the soil core, calculate the oven-dry weight of the soil sample.
5. Calculate the bulk density in g/cm^3 by dividing the weight of soil core, on oven dry basis, by the volume of the soil core, which is the volume of the steel tube.

$$BD\ (g/cm^3) = \frac{Wt.of\ soil\ core\ (dry\ basis)\ (g)}{Vol.of\ soil\ core\ (cm3)}$$

Values of BD of undisturbed cores are of practical significance, as it indicates soil aggregation and structure under field conditions.

e) Measurement of Particle Density

Two methods for the determination of particle density (PD) are discussed. In the first method a graduated cylinder is used, in the second method a volumetric flask is employed. Both analyses are very simple and rapid. The particle density of most soils varies from 2.60 to 2.75 g/cm^3. An average particle density value of 2.67 is commonly used as the specific gravity of soils.

1.The Graduated Cylinder Method

Procedure

1. Weigh 40 g of soil in a 100 ml graduated cylinder.
2. Add 50 ml of water to the soil in the cylinder. Be sure that no soil material is on the inner walls of the cylinder.
3. Stir thoroughly with a stirring rod to displace the air, and rinse the stirring rod and the inner walls of the cylinder with 10 ml water.
4. Allow the mixture to stand for 5 minutes and record the volume of the soil plus 60 ml water.
5. Determine separately the moisture content of the soil sample by the gravimetric method. The amount of moisture should be added to the amount of added water to obtain the total amount of water used.

Calculations

Volume of total solids = Volume of (soil + water) - Volume of added water

Volume of (soil + water) = as shown on graduated cylinder

Volume of added water = 60 ml + soil moisture (ml)

$$\text{Particle Density PD, g/cm}^3 = \frac{\text{Oven-dry wt of soil (g)}}{\text{Volume of total solids (cm3)}}$$

2. The Volumetric Flask Method

Procedure

1. Fill a pre-weighed 100 ml volumetric flask to the mark with boiled distilled water cooled to room temperature.
2. Weigh the flask plus water (W_1), and then discard the water and dry the flask thoroughly.
3. Weigh accurately 50 g of soil and transfer quantitatively to the volumetric flask (W_s).
4. Pipette 50 ml of water into the volumetric flask, washing down any soil particles adhering to the neck.

5. Heat the flask gently on a hot plate until the water starts to boil in order to drive out air from the soil.
6. Cool the content and bring upto the 100 ml mark with boiled distilled water cooled to room temperature.
7. Weigh the volumetric flask with its content (W_2).
8. Determine on a separate sample the moisture content of the soil by the gravimetric method to find the oven-dry weight of the soil in the flask.
9. Calculate first the volume of soil particles and then particle density (PD).

f) Calculations

Volume of soil particles (cm^3) = Ws + (W1-W2)

Particles density (PD), $g/cm^3 = \frac{Ws}{Ws+(W1-W2)}$

Calculation of Percent Pore Space

The pore space (voids) is the portion of bulk soil volume not occupied by solid particles. It is filled with air and water. Depending on pore size, pore spaces are given the name macropores (large) or micropores (small). There is no sharp line of demarcation between the two pore sizes.

Calculations

% pore space (PS) = $\frac{PD-BD}{PD} X\ 100$

❑❑❑

Chapter 11

Chemical Properties

Determination of Soil pH

pH

The soil pH is the negative logarithm of the active hydrogen ion (H^+) concentration in the soil solution. It is the measure of soil sodicity, acidity or neutrality. It is a simple but very important estimation for soils, as soil pH has a considerable influence on the availability of nutrients to crops. It also affects microbial population in soils. Most nutrient elements are available in the pH range of 5.5–6.5. In various chemical estimations, pH regulation is critical. The specific colours as observed in the presence of various pH indicators and the colour changes due to pH change. The apparatus required in order to measure soil pH consists of:

- A pH meter with a range of 0–14 pH;
- A pipette/dispenser;
- Some beakers;
- A glass rod.

The reagents required are:

- Buffer solutions of pH 4, 7 and 9.
- Calcium chloride solution (0.01M): dissolve 14.7 g of $CaCl_2.2H_2O$ in 10 litres of water to obtain 0.01M solution.

The procedure for measuring soil pH is:

1. Calibrate the pH meter, using two buffer solutions, one should be the buffer with neutral pH (7.0) and the other should be chosen based on the range of pH in the soil. Put the buffer solutions in the beakers. Insert the electrode alternately in the beakers containing the two buffer solutions, and adjust the pH. The

instrument indicating pH as per the buffers is ready to test the samples.

Procedure

a) Soil to water ratio of 1:2 (pH)

Take 20 g soil in 100 ml beaker and add 40 ml. of distilled water to it. The suspension is stirred at a regular interval for 30 minutes. Determine the pH by immersing electrodes in suspension. For soils containing high salts, the pH should be determined by using 0.01M calcium chloride solution. (Dissolve 0.110 g of $CaCl_2$ in distilled water).

b) Saturates soil paste (pH)

Add small amount of distilled water to 250g of air dried soil. Stir the mixture with a spatula. At saturation, the soil paste glistens and flows slightly, when the container is tapped it slides freely and ensures cleanly off the spatula. After mixing, allow the sample to stand for an hour. If the paste has stiffened markedly or lost its glistening, add more water or if free water has collected on the surface of the paste, add an additional weighed quantity of dry soil and mix it again. Then insert the electrode carefully in the paste and measure the pH.

c) Saturation extract (pHe)

The soil is extracted using vacuum extractor and the pH is measured in the saturation extract.

Table: pH range and Soil reaction rating

S.No.	pH range	Soil reaction rating
1.	< 4.6	Extremely acidic
2.	4.6–5.5	Strongly acidic
3.	5.6–6.5	Moderately acidic
4.	6.6–6.9	Slightly acidic
5.	7.0	Neutral
6.	7.1–8.5	Moderately alkaline
7.	> 8.5	Strongly alkaline

2. Electrical Conductivity (EC)

The electrical conductivity (EC) is a measure of the ionic transport in a solution between the anode and cathode. This means, the EC is normally considered to be a measurement of the dissolved salts in a solution. Like a metallic conductor, they obey Ohm's law.Since the EC depends on the number of ions in the solution, it is important to know the soil/water ratio used. The EC of a soil is conventionally based on the measurement of the EC in the soil solution extract from a saturated soil paste, as it has been found that the ratio of the soil solution in saturated soil paste is approximately two-three times higher than that at field capacity.

As the determination of EC of soil solution from a saturated soil paste is cumbersome and demands 400-500 g soil sample for the determination, a less complex method is normally used. Generally, a 1:2 soil/water suspension is used.

Apparatus

- EC meter
- Beakers (25 ml), erlenmeyer flasks (250 ml) and pipettes
- Filter paper

Reagent

- 0.01M Potassium chloride solution: Dry a small quantity of AR grade potassium chloride at 60°C for two hours. Weight 0.7456 g of it and dissolve in freshly prepared distilled water and make the volume to one litre. This solution gives an electrical conductivity of 1411.8x10-3 i.e. 1.412 mS/cm at 25°C. For best result, select a conductivity standard (KCl solution) close to the sample value the. The conductivity of soils can above directly known after adjusting the meter with the solution of known electrical conductivity.

Procedure (1)

Take 20 g of soil in 100 ml beaker, add 40 ml of distilled water and shake intermitantly for 30 minutes. Determine the conductivity of the supernatant liquid with the help of conductivity meter. The electrical conductivity of saturation extract (ECe) is also determined for salinity ratings.

EC(dS m^{-1}) Effect

<1 - No deleterious effect on crop 1-2 - Critical for salt sensitive crops
2-3 - Critical for salt tolerant crops >3 - Injurious to most crops

Procedure(2)

1. Take 40 g soil into 250 ml Erlenmeyer flask, add 80 ml of distilled water, stopper the flask and shake on reciprocating shaker for one hour. Filter through Whatman No.1 filter paper. The filtrate is ready for measurement of conductivity.
2. Wash the conductivity electrode with distilled water and rinse with standard KCl solution.
3. Pour some KCl solution into a 25 ml beaker and dip the electrode in the solution.Adjust the conductivity meter to read 1.412 mS/cm, corrected to 25°C.
4. Wash the electrode and dip it in the soil extract.
5. Record the digital display corrected to 25°C. The reading in mS/cm of electrical conductivity is a measure of the soluble salt content in the extract, and an indication of salinity status of this soil (following table). The conductivity can also be expressed as mmhos/cm or dsm^{-1}.

General interpretation of EC values

Soil	EC	Total salt content	Crop reaction
	(mS/cm)	(%)	
Salt free	0–2	< 0.15	Salinity effect negligible, except for more sensitive crops
Slightly saline	4–8 0.	15–0.35	Yield of many crops restricted
Moderately saline	8–15	0.35–0.65	Only tolerant crops yield satisfactorily
Highly saline	> 15	> 0.65	Only very tolerant crops yield Satisfactorily

3. Soil Organic Carbon

Principal

Soil organic matter (SOM) is the seat of nitrogen in soil and its determination is often carried out as an index of nitrogen availability.

Two major methods are generally followed: i) the titration method (Walkley and Black 1934), and ii) the colorimetric method (Datta *et al.*, 1962). In both the methods, organic metter is oxidized with chromic acid (potassium dichromate + H_2SO_4). In the titration method, the unconsumed potassium dichromate is back-titrated against ferrous sulphate or ferrous ammonium sulphate (redox titration). Carbon in the sample is oxidized as follows:

$$2H_2Cr_2O_7 + 3C + 6H_2SO_4 = 3Cr_2(SO_4)_3 + 3CO_2 + 8H_2O$$

Thus, $2H_2Cr_2O_7$ or $2K_2Cr_2O_7 = 3C$

Or 588 g of $K_2Cr_2O_7$ = 36 g of C

Or 12 litres of 1*N* $K_2Cr_2O_7$ = 36 g of C

Or 1 mL of 1*N* $K_2Cr_2O_7$ = 0.003 g of C

In the colorimetric method, the intensity of the green colour of chromic acid obtained due to its reduction is measured calorimetrically. This intensity is directly proportional to the amount of organic carbon present in the soil.

Walkley and Black Method

Reagent

1. **1*N* Potassium dichromate:** Dissolve 49.04 g of AR grade $K_2Cr_2O_7$ in about 500 mL of distilled water and make the volume to one litre.
2. Conc. Sulphuric acid
3. **0.5 N Ferrous ammonium sulphate:** Dissolve 196 g of ferrous ammonium sulphate in distilled water, add 20 mL of conc. H_2SO_4 and make volume to one litre. The ferrous ammonium sulphate should be from a fresh lot and light green in colour. Yellowing of the salt indicates its oxidation.
4. **Diphenylamine indicator:** Dissolve 0.5 g of the dye in a mixture of 20 mL of distilled water and 100 mL of conc. H_2SO_4.
5. Orthophosphoric acid (85%) or sodium fluoride

Procedure

1. Weigh 1 g of 0.2 mm soil sample into 500 ml dry conical flask of borosilicate glass e.g. Corning, Borosil, etc.

2. Add 10 ml of 1*N* $K_2Cr_2O_7$ and 20 ml of conc. H_2SO_4.
3. Swirl a little and keep on an asbestos sheet for 30 minutes.
4. Add slowly 200 ml of distilled water and 10 ml of orthophosphoric acid.
5. Add 1 mL of diohenylamine indicator.
6. Take 0.5 *N* ferrous ammonium sulphate solution in 50 mL burette.
7. Titrate the contents until green colour starts appearing.
8. If the titre value is <6, repeat taking 0.2 to 0.5 g of soil sample.
9. Note:
   i) Use of NaF along with H_2PO_4 gives a sharper end point.
   ii) High chloride content, as in case of saline soils, interferes in the estimation. It can be prevented by adding Ag_2SO_4 @ 1.25 per cent to the conc. sulphuric acid.

Calculations

Organic carbon (%) in soil = 10(B-S)/Bx0.003x100/wt. of sample (g)

Where, B and S stand for the titer values (mL) of blank and sample, repectively.

a) Colorimetric method

Instrument

Colorimeter or Spectrophotometer.

Reagents

1. **1*N* Potassium dichromate:** Dissolve 49.04 g of $K_2Cr_2O_7$ in distilled water and make to 1L.
2. Conc. sulphuric acid
3. Anhydrous sucross (AR grade)

Procedure

1. Weigh 1 g of 0.2 mm soil sample into a 100 mL conical flask.
2. Add 10 mL of 1*N* $K_2Cr_2O_7$ and 20 mL of conc. H_2SO_4.
3. Swirl a little and keep on an asbestos sheet for 30 minutes.

4. Decant about 15 to 20 mL of the supernatant liquid carefully into a centrifuge tube, retaining soil particles in the flask as far as possible.
5. Centrifuge for about 5 minutes to ensure the complete settling of soil particles, if any.
6. Carefully decant the clear liquid into the colorimeter/ spectrophotometer tube and read the colour intensity using red filter/660 nm wavelength.

Simultaneously run a blank without soil and a series of standards as given on the next page.

1. Accurately weight 5, 10, 15, 20 and 25 mg of anhydrous sucrose crystals into 100 mL conical flasks.
2. Proceed as above for oxidation of sucrose and colour intensity measurement.
3. Plot the colorimeter/spectrophotometer readings against quantity of sucrose or carbon.
4. Find out a factor to be multiplied by sample reading to get organic carbon content (%) of soil. If the standard curve is prepared accurately and repeatedly, the average value of this factor comes to be 0.0042 with little possible variation.

Calculations

Organic carbon (%) = colorimeter reading x 0.0042

Caution: The centrifuge may get damaged due to accidental spilling of chromic acid during centrifuging the contents. Keeping the contents overnight is suggested as an alternative. Appearance of chromic acid crystals is sometimes observed even before 30 minutes of keeping for oxidation. Dilution of the contents by adding 10 mL of distilled water after 30 minutes reaction, followed by keeping overnight is useful, if the soil texture is light. For fine-textured soils it is, however, necessary to centrifuge.

4. Soil Organic Matter

Walkley – Black Wet Combustion Method

Soil organic matter is defined as the organic fraction of soil including plant, animal and microbial residues at all stages of

milliequivalent weight of carbon, by 89 and multiplying by 100 to convert to per cent. Chloride interference is eliminated by the addition of the silver sulphate to the digesting acid as indicated. The presence of nitrates and carbonates up to 5 per cent and 50 per cent, respectively, do not interfere.

5. Determination of Soluble Sodium and Potassium

Apparatus

- Flame photometer

Reagents

1. Standard K and Na solutions: Prepare a series of K and Na standard solutions in the range of 0-2 meq/l of K, and 0-4 meq/l of Na. For better results, add 212.5 mg LiCl in each standard to yield a final concentration of about 5 meq/l of LiCl. For the preparation of 2 meq K/l, dissolve 149.2 mg KCl in a litre of distilled water and for the preparation of 4 meq Na/l, dissolve 234 mg NaCl in a litre of distilled water.

Procedure

1. Switch on the flame photometer and let it warm-up for 15 minutes.
2. Calibrate the instrument with blank and standard solutions.
3. Fill the capsules with the soil extracts. Insert the suction tubing in the capsule and record the reading. Dip the tubing in distilled water to wash the system, and then read the sample.
4. The readings for samples express the concentration of K or Na as meq/l of the saturation extract.

Calculations

$$\text{meq of soluble Na or K/100g soil} = \frac{meq/lit\ X SP}{1000}$$

6. Determination of Soluble Calcium and Magnesium

Water-soluble calcium and magnesium are measured in the soil saturation extract. Exchangeable Ca and Mg are determined in ammonium acetate extraction solution. Concentration of Ca and of

Mg can be determined by atomic absorption methods are used because these cations are present in high concentrations in arid and semi-arid soils. To use an atomic absorption spectrometer, a number of dilutions are required.

Apparatus

- Microburette - 10 ml

Reagents

1. Ammonium chloride - Ammonium hydroxide buffer solution: Dissolve 67.5 g of ammonium chloride in 570 ml of concentrated ammonium hydroxide and dilute to 1 litre.
2. Sodium hydroxide (NaOH) 4 M: Dissolve 160 g of sodium hydroxide in 1 litre distilled water.
3. Calcium chloride ($CaCl_2$) standard 0.01 N: Dissolve 0.5 g of pure calcium carbonate crystals in 10 ml of approximately 3 M hydrochloric acid and dilute to 1 litre.
4. Eriochrome black T indicator: Dissolve 0.5 g of Eriochrome black T and 4.5 g of ydroxylamine hydrochloride in 100 ml of 95% ethanol.
5. Calred indicator, 2-Hydroxy-1-(2-Hydroxy-4 Sulfo-1-Naphthyle 20)-3-Naphtholic acid-original salt.
6. Ethylene diaminetetraacetic acid (EDTA) 0.01 N: Dissolve 2.0 g of EDTA in water and make up to 1 litre. Standardize the solution against 0.005 M standard $CaCl_2$ solution.

Procedure

a) Determination of calcium

   1. Pipette an aliquot (2 to 5 ml) of soil extract in 50 ml white porcelain dish.
   2. Dilute with distilled water to a volume of approximately 25 ml.
   3. Add 2 ml of 4 M NaOH and 2-3 mg of calred indicator.
   4. Titrate the contents slowly with 0.01 N EDTA until a sky-blue endpoint is obtained.

5. If the sample is over titrated with EDTA, it can be back titrated with the standard 0.01 N calcium chloride solution.
6. Prepare a blank using 2-5 ml distilled water, and follow steps 2 to 4.
7. Express the Ca in meq/litre of the saturation extract.

Calculations

Meq of Ca/litre

(meq EDTA used for soil extract - me EDTA for blank)

$$1000 \times \frac{\text{mg of EDTA solution used} + \text{N of EDTA}}{\text{mg of solution taken (aliquat)}}$$

b) Determination of Calcium and Magnesium

1. Pipette an aliquot (2-5 ml) into 50 ml porcelain evaporating dish.
2. Dilute with distilled water to a volume of about 25 ml and add 5 ml of ammonium chloride - ammonium hydroxide buffer solution and 3 to 4 drops of Eriochrome black T indicator.
3. Titrate the contents with 0.01 N EDTA until a sky-blue end point is obtained.
4. Prepare a blank using 2–5 ml distilled water, and follow steps 2 and 3.
5. Express the amount of Ca + Mg in the sample in meq/l.

Calculations

(Ca+Mg) meq/litre

(me EDTA used for soil extract – me EDTA for blank)

$$1000 \times \frac{\text{mg of EDTA solution used} + \text{N of EDTA}}{\text{mg of solution taken (aliquat)}}$$

Determination of Magnesium

Concentration of Mg, as meq/l, is calculated by subtracting meq/l of Ca from meq/l of (Ca + Mg). The difference is the concentration of Mg as meq/l.

7. Determination of Carbonate and Bicarbonate

Carbonate and bicarbonate ions are species of the same acid, carbonic acid. Their proportionate content is a function of pH. The CO_3^{2-} starts to form as pH rises above 8.4.

Apparatus

- Magnetic stirrer
- Microburette - 10 ml
- White Porcelain crucibles

Reagents

1. Sulphuric acid (H_2SO_4) standard solution, 0.01 M. Dilute 0.56 ml of conc. sulphuric acid in 1 litre distilled water and standardize using a primary standard to determine exact molarity.
2. Phenophthalein indicator: Dissolve 0.25 g of phenolphthalein in 100 ml of 60% alcohol.
3. Methyl orange indicator: Dissolve 0.1 g of methyl orange in 100 ml of water.

Procedure

1. Pipette 5 ml of extract into 50 ml Erlenmeyer flask and dilute with boiled distilled water to a volume of approximately 25 ml.
2. Add 3 or 4 drops of phenolphthalein indicator. The appearance of a pink colour indicates the presence of carbonates in the sample.
3. Place the flask on a magnetic stirrer and titrate the content in the flask with 0.01 M H_2SO_4 by adding one drop every 2-3 seconds until the pink colour disappears.
4. Record the volume of H_2SO_4 titrant used (Vph).
5. To the colourless solution add 2 or 3 drops of methyl orange indicator.
6. Continue the titration, without refilling the burette, to the pink endpoint.

7. Record the total volume of H_2SO_4 used (Vt). Save the solution for chloride determination. Make a blank correction for the methyl orange titration. Express the amount of CO_3^{2-} and HCO_3^- in the sample as meq/l.

Calculations

a) For CO_3^{2-} concentration, the phenolpthalan endpoint is considered

$$\text{Meq/litre of } CO_3^{2-} = \frac{1000\ ml}{5\ ml\ aliquat} X\ 2\ (Vph\ x\ M\ of\ H2SO4)$$

b) $$\text{Meq/litre } HCO_3 = \frac{1000\ ml}{5\ ml\ aliquot} X\ M\ of\ H2SO4\ (Vt - 2Vph)$$

8. Determination of Chloride

Chloride is usually determined in soil saturation extract to assess the concentration of soluble salts. Estimation by silver nitrate tiration method is described.

Apparatus

- 10 ml microburette

Reagents

1. Potassium Chromate (K_2CrO_4) indicator 5%: Dissolve 5 g of K_2CrO_4 into 90 ml water. Add a saturated $AgNO_3$ solution until some brownish red $AgCrO_4$ precipitate forms. Place the solution in dark for 24 hours, filter and make the volume to 100 ml.
2. Silver nitrate ($AgNO_3$) standard, 0.005 M: Dissolve 0.8495 g of $AgNO_3$ in distilled water and dilute to 1 litre.

Procedure

1. Add 5 drops of K_2CrO_4 indicator to the solution saved after the titration of the carbonate and bicarbonate, or start with another 5 ml aliquot of the extract.
2. Titrate the contents under bright light with the standard $AgNO_3$ to a brownish-reddish endpoint.

3. Prepare a blank in order to (a) correct for the amount of Ag^{2+} used to form the silver chromate (Ag_2CrO_4) red precipitate, and (b) to use as a reference for the end point. The volume of solution at the end of titration of the blank should be almost equal to that of the unknown.

Calculations

Meq of Cl/litre = $\frac{1000\ ml}{5\ ml\ aliquot} X\ ml\ of\ AgNO3\ used$ X 0.005 M of $AgNO_3$

9. Determination of Calcium Carbonate

Calcium carbonate, a major component of calcareous soils, ranges from a few percent in slightly calcareous soils to more than 80% in some extremely calcareous soils. The amount and form of calcium carbonate (total versus active) and its distribution down the soil profile, affect soil physical and chemical characteristics. It was found that when total $CaCO_3$ was above 20%, active $CaCO_3$ was more than 10%. Carter (1981) gave 11 to 30% total carbonate and 7-9% active $CaCO_3$ as the critical levels that adversely affected crop growth.

The authors showed that soil retention of P, Mn, Zn and Cu was directly related to carbonate content and to the distribution of total and active calcium carbonate between the clay and silt fractions. When total $CaCO_3$ was less than 20%, the retention of those elements was affected mainly by the total amount of carbonates, but, when it was above 20%, nature of the carbonates was more important in governing retention. The active $CaCO_3$ was more than 10% when the total $CaCO_3$ was above 25% in the studied soils. Therefore, one could give the limit of about 10% active $CaCO_3$, which will contribute significantly to the retention of certain essential lements in calcareous soils.

Physical effects: The active-$CaCO_3$ portion and the distribution of $CaCO_3$ in the different particle size fraction affects the properties (soil-water relationships) of calcareous soil (Deb and Chadha, 1970; Thabet, 1975). Calcareous soils with 15% calcium carbonate content have higher water diffusivity and faster water movement than non-calcareous soils of similar texture. Evidently, $CaCO_3$ assists in the formation of stable soil aggregates. But, increased $CaCO_3$ content (25% or more) tends to precipitate within the capillary tubes causing an increase in the proportion of micropores, thus reducing the percolation of water.

Effect on plants: Continuous horizons of carbonate accumulation in soil profile may prevent root penetration and thus, may retard the degree of plant development and reduce the yield. Sys (1975) classified the sensitivity of crops to $CaCO_3$ into three groups as follows:

1. Tolerant crops: wheat, alfalfa, figs, olives and dates
2. Moderately tolerant crops: barley, clover, cotton, maize, millets, rice,grapes, sugarcane, sugar beet, watermelons, lettuce, tomatoes, beans, artichokes, tobacco and onions.
3. Sensitive crops: citrus, banana and potatoes.

8. Determination of Total Calcium Carbonate

Apparatus

- 25 ml and 100 ml Erlenmeyer flasks
- Burette, 25 ml

Reagents

1. Hydrochloric acid (HCl) Standard 1 M: Add 81 ml HCl to about 500 ml water in a 1 litre volumetric flask, cool and make to volume with distilled water. Standardize it against a primary standard.
2. Sodium hydroxide (NaOH) Standard 0.5 M: Dissolve 20 g NaOH in 800 ml distilled water in a 1 litre volumetric flask, cool and dilute to 1 litre. Standardize it against a primary standard.
3. Phenolphthalein indicator 1% in 60% ethanol: Weigh 1 g phenolphthalein crystals in 100 ml volumetric flask add 60 ml of ethyl alcohol and dilute to volume with distilled water. Shake the flask until the crystals dissolve completely.

Procedure

1. Weigh 5 g of soil and transfer into a 250 ml Erlenmeyer flask.
2. Add 100 ml of 1.0 M HCl.
3. Cover the flask with aluminium foil and keep it overnight or heat it to boiling for 5 minutes and let it cool to room temperature.
4. Filter, and pipette 10 ml of the filtrate into a 100 ml Erlenmeyer flask.

5. Add 2 or 3 drops of phenolphthalein to the content and titrate with 0.5 M sodium hydroxide.

Calculations

$\%CaCO_3$ = [(ml Hcl x 1M) - (Me NaOH x0.5M)] x

$$\frac{vol.HCl\ 100\ ml}{vol.filterate\ aliquot\ (10\ ml)} X \frac{100}{1000x2} X \frac{100g}{wt\ of\ soil\ g}$$

% $CaCO_3$ = [(10 x 1) - (ml NaOH x 0.5)] x 10

10. Cation exchange capacity CEC

Presence of high concentrations of calcium, especially in the form of gypsum interferes with the determination of cation exchange capacity, which is an important parameter for soil fertility and mineralogical characterization.

The surface of soil particles are expressed in terms of milliequivalents of negative charge per 100 g of oven-dried soil (1 meq/ 100 g soil = 1 cmol/kg in SI system). The unit meq/100 g has been used in this manual because most of the soil testing laboratories in the Near East still use and are familiar with meq/100g for CEC and exchangeable cations measurements.

a) Determination of Cation Exchange Capacity (CEC)

Apparatus

- Centrifuge
- 50 ml round bottom centrifuge tubes
- Mechanical shaker
- Flame photometer

Reagents

1. Sodium acetate (NaOAc) 1.0 M: Dissolve 136.08 g of sodium acetate trihydrate in distilled water and bring volume to 1 litre.Adjust the pH to 8.2.
2. Ethanol 95% addition of more ammonium hydroxide or acetic acid.
3. Na standard solution.

Procedure

1. Weigh accurately about 5 g soil and transfer the sample to a 50 ml centrifuge tube.
2. Add 30 ml of 1.0 M sodium acetate solution to the tube, stopper and shake in a mechanical shaker for 5 minutes.
3. Centrifuge at 2000 rpm for 5 minutes or until the supernatant liquid is clear.
4. Decant the liquid completely and repeat the extraction three more times.Discard the decants.
5. Repeat steps 2-4 with ethanol or isopropyl alcohol until the EC of the decant reads less than 40 mS/cm (usually it takes 4 to 5 washings).
6. To displace the adsorbed Na, repeat steps 2-4 using the ammonium acetate solution. Collect the decants in 100 ml volumetric flask fitted with a funnel and filter paper. Make up to volume with ammonium acetate solution.
7. To determine sodium concentration by flame photometry, prepare a series of Na standard solutions in the range of 0-4 meq/l of Na.

Calculations

Meq of Na/100 gm soil emission (R meq/l) 100 ml

$$= \frac{R X 10}{wt\ of\ soil\ (g)} X \frac{10}{1000\ ml} X\ 100\ ml$$

Where, R is the meq/l of Na as determined by the flame photometer. The displaced Na is actually a measure of the Cation Exchange Capacity (CEC) of the soil. So, the meq/100 g of Na is actually meq/100 g exchangeable cations (Ca, Mg, Na and K).

b) Determination of CEC in Gypsiferous and Calcareous Soils

Apparatus

- Centrifuge
- 50 ml round bottom centrifuge tubes
- Mechanical shaker
- Flame photometer

Reagents

1. Sodium oxalate ($Na_2C_2O_4$) saturated solution: Add 10 g sodium oxalate crystals to 100 ml distilled water. Allow to stand with occasional shaking for 8 hrs or longer. Let it settle until clear or filter.
2. Ethanol 95% or isopropyl alcohol 3. Ammonium acetate (NH_4OAc) 1.0 M solution: Add 57 ml of conc. Acetic acid (reagent grade) to 700-800 ml distilled water in a 1 litre beaker. Then, with constant stirring, add 68 ml of conc. ammonium hydroxide. Bring the volume close to 1 litre, and adjust pH to 7.0 by the addition of more ammonium hydroxide or acetic acid.
3. Or, dissolve 77.09 g of ammonium acetate in distilled water and dilute to 1 litre. Adjust pH to 7.0.
4. Na standard solution

Procedure

1. Weigh accurately 4 g of soil and transfer to a 50 ml centrifuge tube.
2. Add 30 ml of saturated sodium oxalate solution to the tube, stopper and shake for two hours.
3. Centrifuge at 2000 rpm for 5 minutes or until the supernatant liquid is clear.
4. Decant the liquid completely and repeat the extraction three more times Discard the decants.
5. Repeat steps 2-4 with ethanol 95% or isopropyl alcohol until the EC of the decant reads less than 40 mS/cm (takes 4 to 5 washings).
6. To displace adsorbed sodium, repeat steps 2-4 using ammonium acetate solution. Collect the decants in a 100 ml volumetric flask fitted with a funnel and filter paper. Make up to volume with ammonium acetate solution.
7. Determine sodium concentration

Calculations

$$\text{Wt. of soil (g)} = \frac{R X 10}{wt\ of\ soil\ (g)} X \frac{100\ g}{1000\ ml} X\ 100\ ml$$

meq of Na/100 g soil Emission Reading (R meq/l) X 100 ml

Where R is the meq/l of Na as determined by the flame photometer. As was explained above (section 8.1), meq/100 g of Na displaced by NH4, is actually the meq/100 g of exchangeable cations (Ca, Mg, Na and K) or the CEC of the soil.

11. Exchangeable Sodium Percentage

Exchangeable Sodium Percentage (ESP) is the relative amount of exchangeable Na to the sum of exchangeable cations, or CEC.

As more and more of the cation exchange capacity of soil becomes satisfied with Na on the expense of Ca, Mg and K, the clay fraction become dispersed and deflocculated causing the destruction of soil structure. Some critical values were internationally given to ESP for the classification of sodium-affected soils and its effect on soil structure and plant growth. Ammonium acetate extract is analysed for determination of Exch. Na and K.

Calculations

$$\text{ESP (Exch. Sodium Percentage)} = \frac{\text{Na Conc. in amm. acetate extract}}{\text{CEC}} \times 100$$

Reagent

Acetone (analytical grade)

Procedure

1. Place 1 g of air-dried soil, ground to pass through a 270 mesh (0.053 mm) sieve, in an Erlenmeyer flask.
2. Add 500 ml of distilled water, stopper the flask, and shake for 20 minutes on a mechanical shaker.
3. Filter the content through Whatman No. 41 filter paper.
4. Transfer 20 ml aliquot of the filtered extract into a 50 ml centrifuge tube.
5. Add 20 ml of acetone, mix and allow to stand for 15 minutes or until the precipitate locculates.

6. Centrifuge the content at 2000 rpm for 3 minutes, carefully decant the supernatant, then invert the tube on a clean filter paper and let it drain for 5 minutes.
7. Wash the sides of the tube with 10 ml acetone and disperse the precipitate.
8. Repeat step 6.
9. Add 40 ml of distilled water to the tube, stopper and shake until the precipitate is completely dissolved.
10. Measure the EC of the solution and correct conductivity reading to 25°C.
11. Determine the gypsum concentration in solution chart from which shows a direct relationship between the EC of the extract and its gypsum content.

❑❑❑

Chapter 12

Major Soil Nutrients

1. Soil Nitrogen

Total N includes all forms of inorganic N, like NH_4-N, NO_3-N and also NH_2 (Urea)-N, and the organic N compounds like proteins, amino acids and other derivatives. Depending upon the form of N present in a particular sample, specific method is to be adopted for getting the total nitrogen value. While the organic N materials can be converted into simple inorganic ammoniacal salt by digestion with sulphuric acid, for reducing nitrates into ammoniacal form, use of salicylic acid or Devarda's alloy is made in the modified Kjeldahl method. At the end of digestion, all organic and inorganic salts are converted into ammonium form which is distilled and estimated by using standard acid. As the precision of the method depends upon complete conversion of organic N into NH4 - N, the digestion temperature and time, solid:acid ratio and the type of catalyst used have an important bearing on the method. The ideal temperature for digestion is 32°- 37°C. At lower temperature, the digestion may not be complete, while above 41°C, the loss of NH_3 may occur. The salt: acid (weight: volume) ratio should not be less than 1:1 at the end of digestion. Commonly used catalysts to hasten the digestion process are $CuSO_4$ or Hg. Potassium sulphate is added to raise the boiling point of the acid so that loss of acid by volatilization is prevented.

Total nitrogen is estimated by the micro-Kjeldahl method as per procedure suggested by AOAC (1995).

Preparation of soil samples

The soil sample from definite depth was randomly collected from the field with the help of screw auger. All the possible technical precautions as prescribed for standard soil sampling were also taken. Samples were brought to the laboratory, air-dried in the shade and grounded by wooden roller, thereafter sieved through 2 mm stainless steel sieve and stored in polythene bags and used for chemical assay.

Principle

Nitrogen in samples like soil exists in a very complicated bonding structure. During digestion, a known weight of the soil samples in the presence of sulphuric acid with catalyst mixture under high temperature is digested where complicated structures are broken to simple structure, thereby releasing nitrogen in the form of ammonium radical (NH_4^+). During distillation in presence of sodium hydroxide, the released ammonia is condensed and absorbed in known volume of a boric acid with mix indicator to form ammonium borate, the excess of which is titrated with a standard sulphuric acid.

The micro-Kjeldahl method consists of the three steps :

1. Digestion
2. Distillation and
3. Titration.

Equipment and apparatus

Through KEL PLUS Automatic Nitrogen Estimation System : The said instrument is used for determination of nitrogen. It consists of the following :

a) **Macro Block Digestion System (Model KES 12L):** This digestion system is suitable for soil, plant, water, pesticides, fertilizers, food and feed samples. It is microprocessor based automatic twelve place macro block digestion system with temperature controller fitted with sensor break protection (Microprocessor based) feature and temperature range from 50-450°C.

b) **Acid Neutralizer Scrubber (Model KEL VAC):** It is used to neutralize the acid fumes, which are absorbed in 15% sodium hydroxide and dissolved in water stored in the system tank. After every 2 cycles of digestion, the 15% sodium hydroxide solution is replaced and after 3 cycles of digestion, acid fumes dissolved in water tank is drained off and refilled with fresh water in the system tank.

c) **Automatic Distillation System (Model Classic DX):** It is fully automatic distillation system with programmable auto run digital features, with automatic dilution and addition of boric acid and NaOH. Both modes (auto and manual) are available for distillation reagents addition.

d) **Refrigerated Water Cooling System for Condenser (Model Kel Freeze):** It is refrigerated water cooling system for distillation and condensing system with inbuilt compressor and recirculator pump.

1. Electronic balance
2. Burette
3. Pipette
4. Conical flask
5. Measuring cylinder
6. Distilled water

Reagents

1. Concentrated sulphuric acid (H_2SO_4).
2. Catalyst mixture: Mix with 250 g potassium sulphate (K_2SO_4), 50 g cupric sulphate ($CuSO_4$. 5 H_2O) and 5 g metallic selenium powder in the ratio of 50:10:1.
3. 40 % sodium hydroxide (NaOH).
4. 4 % boric acid containing 20 - 25 ml mixed indicator/liter.
5. Mixed indicator: 0.066 g methyl red + 0.099 g bromocresol green dissolve in 100 ml of 95 % alcohol.
6. 0.02N sulphuric acid (H_2SO_4).

Procedure

I. Digestion

i) Weigh 1 g of soil sample and transfer it to the digestion tube.

ii) Add 10 ml of concentrated sulphuric acid and 5 g of catalyst mixture to the sample.

iii) Load the digestion tubes in to the digester and heat the digestion block.

iv) Switch on the digestion unit and set the initial temperature 100 ^{0}C till frothing is over.

v) Then block temperature is raised to 400°C. The effective digestion starts only at 360°C and beyond 410°C.

vi) The sample turns light green colour or colorless at the end of the digestion process.

II Distillation

i) After cooling the digestion tube, load the tube in distillation unit and other side of hose keep 20 ml of 4 % boric acid with mixed indicator in 250 ml conical flask.

ii) 40 ml NaOH (40 %) is automatically added by distillation unit programme.

iii) The digested sample is heated by passing steam at a steady rate and the liberated ammonia absorbed in 20 ml of 4 % boric acid containing mixed indicator solution kept in a 250 ml conical flask.

iv) With the absorption of ammonia, the pinkish colour turns to green.

v) Nearly 150 ml of distillate is collected in about 8 minutes.

vi) Simultaneously, blank sample (without plant/soil) is to be run.

III) Titration

a) The green colour distillate is titrating with 0.02N sulphuric acid and the colour changes to original shade (pinkish colour).

b) Note the blank and sample titer reading (ml) and calculate the total nitrogen content present in soil samples.

Calculations

R (sample titer-blank titer) x

$$\textbf{Nitrogen content in soil} = \frac{\text{Normality of acid x Atomic Weight of Nitrogen}}{\text{Sample weight (g) x 1000}}$$

$$= \frac{R \times 0.1 \times 14 \times 100}{0.5 \times 1000}$$

Factor = R x 0.14

Crude protein content : The total nitrogen is estimated by micro-Kjeldahl method as per procedure suggested by AOAC (1995) and the crude protein is calculated by the following formula:

Crude protein content (%) = micro-Kjeldahl nitrogen content (%) x 6.25 (based on the assumptions that nitrogen constitutes 16 % of protein).

3. Determination of Available (Mineralization) Nitrogen

(Alkaline Permanganate Method)

Preparation of soil sample : The soil samples from definite depth are randomly collected from the field with the help of screw auger. All the possible technical precautions as prescribed for standard soil sampling are also taken. Samples are then brought to the laboratory, air-dried in the shade and grounded by wooden roller, thereafter sieved through 2 mm stainless steel sieve and stored in polythene bags and used for chemical assay.

Principle : A known weight of the soil is mixed with alkaline potassium permanganate ($KMnO_4$) solution and distilled. The organic matter present in soil is oxidized by the nascent oxygen, liberated by potassium permanganate, in the presence of sodium hydroxide and the released ammonia is condensed and absorbed in known volume of a boric acid with mix indicator to form ammonium borate, the excess of which is titrated with a standard sulphuric acid.

Reactions involved

I) Distillation

Alkaline

$$2K\,M_nO_4 + H_2O \xrightarrow[\text{Medium}]{\text{Alkaline}} 2MnOH + 2KOH + 3O$$

(Nascent oxygen)

$$R.CHNH_2\,COOH + O^- \xrightarrow{\text{Oxidation}} R.CO.COOH + NH_3$$

(Ammonia)

Organic - N fraction

$$NH_3 + H_2O \xrightarrow{\text{Distillation}} NH_4\,OH$$

$$3NH_4\,OH + H_3\,BO_3 \xrightarrow{\text{Absorption}} (NH_4)_2\,BO_3 + 3H_2O$$

II) Titration

$$2(NH_4)_3\,BO_3 + 3H_2SO_4 \xrightarrow{\text{Green colour}} 2(NH_4)_3\,SO_4 + 2H_3\,BO_3$$

(Pink Colour in Original)

I. Distillation

Equipment and apparatus

1. KEL PLUS Automatic Nitrogen Estimation System

The said instrument is used for determination of available nitrogen in soil. It consists of the following:

Principle

- A known weight of the soil is mixed with alkaline potassium permanganate ($KMnO_4$) solution and distilled. The organic matter present in soil is oxidized by the nascent oxygen, liberated by potassium permanganate, in the presence of sodium hydroxide and the released ammonia is condensed and absorbed in known volume of a boric acid with mix indicator to form ammonium borate, the excess of which is titrated with a standard sulphuric acid.
- **Automatic Distillation System (Model Classic DX):** It is fully automatic distillation system with programmable auto run digital features, with automatic dilution and addition of boric acid, NaOH and $KMnO_4$. Both modes (auto and manual) are available for distillation reagents addition.
- **Refrigerated Water Cooling**
- **System for Condenser (Model Kel Freeze):** It is refrigerated water cooling system for distillation and condensing system with inbuilt compressor and recirculator pump.
- Electronic balance
- Burette
- Conical flask
- Distilled water

Reagents

1. 0.32 % potassium permanganate ($KMnO_4$) solution.
2. 2.5 % sodium hydroxide (NaOH).
3. 2 % boric acid solution containing 20-25 ml of mixed indicator/ litre.

4. Mixed indicator: 0.066g methyl red + 0.099g bromocresol green dissolve in 100 ml of 95 % alcohol.
5. 0.02 N sulphuric acid (H_2SO_4).

Procedure

- Weigh 5 g of prepared soil sample and transfer it to the digestion tube.
- Load the tube in distillation unit and other sides of hose keep 20 ml of 2 % boric acid with mixed indicator in 250 ml conical flask.
- 25 ml each of potassium permanganate (0.32 %) and sodium hydroxide (2.5 %) solution is automatically added by distillation unit programme.
- The sample is heated by passing steam at a steady rate and the liberated ammonia absorbed in 20 ml of 2%boric acid containing mixed indicater solution kept in 250 ml conical flask.
- With the absorption of ammonia, the pinkish colour turns to green.
- Nearly 150 ml of distillate is collected in about 10 minutes.
- The green colour distillate is titrating with 0.02N sulphuric acid and the colour changes to original shade (pinkish colour).
- Simultaneously, blank sample (without soil) is to be run.
- Note the blank & sample titer reading (ml) and calculate the available nitrogen in soil.

Calculations

$$\text{Avaiable N Kg ha}^{-1} = \frac{\text{RC Titer reading - x Normality Blank reading of acid X Atomic Weight of Nitrogen x Wt of one hectare of Soil}}{\text{Sample weight (g) x 1000}}$$

$$= \frac{R \times 0.02 \times 14 \times 2.24 \times 10^6}{5 \times 1000}$$

Factor = R x 125.44

Interpretation of results

Available N (kg ha^{-1})		Soil rating
< 280	:	Low
280-560	:	Medium
> 560	:	High

4. Soil Phosphorus

In general, arid and semi-arid soils are low in available phosphorus and plants respond to application of phosphatic fertilizers. Applied fertilizer phosphate quickly reverts to insoluble forms. The amounts of the various discrete chemical fractions of P, which are formed in soil determine the relative effectiveness of phosphatic fertilizers on crop growth (Lindsay and De Ment, 1961), and are related to the genesis of soils (Westin and Buntley (1966). The retention of phosphorus in soils is the result of chemical precipitation and physico-chemical adsorption (Hemwall, 1957). Bashour and Al-Jaloud (2000) studied the fate of applied phosphorus in arid calcareous soils of Saudi Arabia and found that 1/3 of the applied P fertilizers was taken up by plants and about 1/3 was retained in the soil as calcium phosphate. Most of the remaining 1/3 was converted into organic-P and $NaHCO_3$ extractable-P. Sayegh and Abdul Majid (1969) studied phosphorus fractionation and retention in soils of the semi-arid region, and the results are briefly indicated below.

4.1. Fractionation of Soil Phosphorus

Phosphorus characterization indicated that the water-soluble and easily replaceable P increased while Al-P, Fe-P, reductant soluble Fe-P, and occluded Al-Fe-P decreased with an increase of $CaCO_3$ content. The Ca-P in the soils increased with the increase of $CaCO_3$ content, amounting to 88 per cent of the total phosphorus in highly calcareous soils. In the alkaline low calcareous soils, Ca-P was the least, amounting to 5 percent, while Fe-P amounted to 19 per cent of the total phosphorus suggesting a comparatively high iron content which is supported by the soil analysis and by the dark reddish brown colour. Thus, the considerable differences in the proportions of the various P fractions were associated with the degree of calcareousness.

4.2. Fate of Applied Phosphatic Fertilizer

The applied phosphorus was mainly converted to the forms of Ca-P, water soluble P and easily replaceable P in the highly calcareous soils. In the slightly calcareous soils, the distribution of the applied P was mainly in the forms of Ca-P and Al-P, followed by water soluble P. In contrast to the calcareous soils, in the alkaline non-calcareous soils with iron and aluminium oxides, the applied P was distributed in a variety of forms decreasing in the following order: residual-P > Fe- P > Al-P > Ca-P> organic

4.3. Influence of Soil Parent Materials on the Distribution of *the Inorganic P-fractions*

The highly calcareous soils had the highest percentage of Ca-P, and no Fe-P, reluctant soluble Fe-P, and occluded Al-Fe-P. In contrast, the soil developed from the hard Eocene calcareous rock had a greater percentage of all the P fractions except for the Ca-P. The soil derived from hard siliceous calcareous beds contained intermediate amounts of P fractions.

4.4. Phosphorus Retention

Phosphorus retention of the soils increased with the increase of iron and aluminium oxides. The retention of P was more rapid in the alkaline non calcareous soils, which contained larger amounts of iron and aluminium oxides than in the low and high calcareous soils. The slower retention in the calcareous soils might be due to the continued slow precipitation of calcium phosphates in the course of time. The removal of the amorphous materials from the clay fractions of the low and high calcareous soils increased P retention from 2 to 70 per cent upon the addition of 250 μg/g of P, in the highly calcareous soils. The slight difference in the cation exchange capacity between the clays without any amorphous material or clays containing little quantity of amorphous material (partially clean and clean clays) did not appear to provide an adequate explanation for the large difference in phosphorus retention. It was, therefore, postulated that the retention capacity of the Ca-saturated samples was due to a surface reaction, thus, the removal of amorphous materials resulted in increased effective surface area and more retention reaction. On the other hand, removal of the amorphous materials from the clay fractions of the non-calcareous soils, which are rich in iron and aluminium oxides, resulted in less P retention. This was attributed to the presence of comparatively high amounts of iron and aluminium in the amorphous materials.

4.5. Iron and Aluminium Amorphous Phosphate in Calcareous Soils

Most text books state that iron phosphates and aluminium phosphates and their intergrades are found in acidic soils, where conditions are favourable for the presence of Al and Fe. While in neutral to alkaline soils, calcium phosphate is present, where conditions are favourable for the presence of calcium. But they fail to mention the importance of iron and aluminium amorphous materials on phosphorus retention, which coat the clays, silt and sand fractions in the neutral and alkaline soils (Sayegh and Abdul Majid, 1969). Tisdale and Nelson (1993) reported that in general the inorganic phosphorus content in arid soils is higher than the organic phosphorus content. Brady (1990) and Sultanpur *et al.* (1988) have reported that the Aridisols and Andisols have 64% and 63% of inorganic phosphorus, respectively.

Estimation of Available Phosphorus

For determining plant available P in soils, two methods are commonly used in India. The Olsen's method (Olsen *et al.*, 1954) is used for neutral-alkaline soils while the Bray and Kurtz P_1 method (Bray and Kurtz 1945) is used for acid soils.

Olsen's ($NaHCO_3$) Method

Sodium bi-carbonate ($NaHCO_3$) solution extracts some exchangeable or surface-adsorbed Al-P, Fe-P, calcium phosphate and other phosphates.

Reagents

1. Sodium bi-carbonate (Olsen's bicarbonate (Olsen's reagent) 0.5 M $NaHCO_3$, pH 8.5. Dissolve 84 g $NaHCO_3$ in water and make up to 2 litre. Mix thoroughly. Adjust to pH 8.5 with 1 M NaOH (4 NaOH/100 ml) solution. Usually 20-25 ml NaOH solution is required for litre $NaHCO_3$ solution. Store in glass or polythlene bottle.

2. Reagent A. Dissolve 12 g ammonium molybdate ($(NH_4)_6Mo_7O_{24}$, $4H_2O$) in 250 ml of distilled water. Dissolve 0.2908 g antimony potassium tartrate ($K(SbO)C_4H_4O_6 1/2H_2O$) in 100 ml water. Add these two solutions to 1000 ml of 2.5 M H_2SO_4, mix thoroughly and make to 2000 ml. Store in pyrex glass bottle in a dark and cool place.
3. Reagent B. Dissolve 1.056 g ascorbic acid ($C_6H_8O_6$) in 200 ml reagent A and mix. This does not keep for more than 24 h at room temperature. Prepare daily as required.
4. Sulphuric acid 2.5 M: Dilute 140 ml of conc. H_2SO_4 to 1 litre.
5. Standard stock P solution : Dissolve exactly 0.439 g potassium dihydrogen orthophosphate (KH_2PO_4) A.R. grade (dried in oven at 60°C for 1 h and cooled in desiccators) in half a litre of distilled water. Add 25 ml of H_2SO_4 (approx.) and make to one litre with distilled water. This gives 100 ppm P standard stock solution. From this, a 2 ppm solution is made by diluting it 50 times.

Standard Curve

To prepare the standard curve, 1, 2, 3, 4, 5 and 10 ml of 2 ppm P solution are taken in 25 ml volumetric flasks. To these 5 ml of the extracting solution (Olsen's or Bray and Kurtz P_1) is added. When Olsen's extractant is used, acidify 5 ml aliquot with 2.5 M H_2SO_4 to pH 5.0. Add distilled water to make the volume to 20 ml and then add 4 ml of reagent B. make the volume with distilled water and mix. Prepare a blank with $NaHCO_3$ solution, distilled water and 4 ml of reagent B. After waiting for 10 minutes, read the intensity of the blue colour in a photoelectric colorimeter using 730-840 μm filter or on a spectrophotometer at 882 μm (Watanabe and Olsen, 1965). Some workers prefer reading at 730 μm o some spectrophotometers as the higher wavelength (882 μm) is not found sensitive.

Procedure

- Weigh 2.5 h 2 mm air-dry soil (0.1 g accuracy) into a 150 mml Erlenmeyer flask.
- Add a little of Darco G 60 or equivalent grade P-free activated charcoal. See later section on suggestions.
- Then add 50 ml of Olsen's reagent (soil to solution ratio of 1:20) and shake on the reciprocating shaker for 30 min (180 + oscillations/min).

- Similarly, run a blank without soil.
- Filter through Whatman no. 40 or 42 filter paper into a clean and dry beaker. Shake the flask immediately before pouring suspension into funnel.
- Place a 5 ml aliquot of the extract in a 25-ml volumetric flask and acidify with 2.5 M H_2SO_4 to pH 5.0. Add distilled water to 20 ml and then add 4 ml of reagent B. After waiting for 10 min, read the intensity of blue color on a spectrophotometer or colorimeter as described for standard curve above.

Calculations

$$\text{Available P (kg/ha)} = \frac{\text{R x Volume of extract}}{\text{Volume of aliquot}} \times \frac{2.24 \text{ x } 10^6}{\text{Wt. (g) of soil x } 10^6}$$

Where R = ug P in the aliquot (obtained from standard curve)

$$\frac{\text{R x 50 x 2.24}}{\text{5 x 2.5}} = \text{ug P x 8.96}$$

Comments and suggestions

Olsen's method is suitable for neutral, alkaline and calcareous soils. It is normally used or alkaline soils of semiarid regions which have high CEC and high base saturation and very often free $CaCO_3$. The extractant ($NaHCO_3$) has two disadvantages. It dissolves organic matter, producing coloured extracts which have to be clarified with Darco-G 6 or activated carbon. The extractant has to be acidified to bring the pH to 5.0 befoure molybdenum blue colour is developed. Neutralisation of bicarbonate with acid produces CO_2 bubbles with interfere with colour development.

These days there is an interest in universal extraction reagents which can be used to determine major elements and micronutrients (Jones 1990). The ammonium bicarbonate-DTPA extractant (AB-DTPA) was introduces for use on alkaline soils (Sultanpur and Schwab 1977). AB-DTPA (1M NH_4HCO_3-0.005 M DTPA: pH 7.6) can substitute for Olsen's method on alkaline soils whose water pH exceeds 7.5 but are not excessive in their Ca or Na content. It can be used as an extractant for K and NO_3-N, i addition to several micronutrients. One of the main problems with this reagent is that it is unstable in its pH and must be made either before use or kept stored under mineral oil.

- To obtain P free Darco-G-60 or activated charcoal, 80 h material as a slurry in distilled water is leached overnight with 1 litre of M HCl in 60 mm diameter columns (250 ml capacity) and then with de-ionized water till the leachate is free of chloride. The material is then dried at 110°C (Sarirullah *et al.*, 1990).
- Before P estimation is carried out, take 5 ml of the extractant in a separate 25 ml beaker and determine the volume of acid required to bring the pH of the solution to 5.0 using p-nitrophenol indicator, the yellow colour of which disappears at this pH. The requisite volume of acid can be added to aliquots and kept aside for sufficient time with occasional shaking by hand to neutralize the bicarbonate and to allow the CO_2 bubbles to escape.
- Prepare a fresh solution of $NaHCO_3$, if solution has been standing over 1 month in a glass bottle. Its pH trends to increase in glass bottle. Use polythene bottle for periods longer than 1 moth but check pH every month (Olsen and Sommers, 1982).
- The ascorbic acid method has proved to be reliable and less subjected to interferences in colour development than the $SnCl_2$ method. The colour is stable for 24 hours.

❐ Temperature and shaking speed increases the P extracted by $NaHCO_3$. Therefore room temperature and shaking speed have to be regulated to obtain precise estimates of the Olsen's P. A shaking time of 30 minutes, and speed of 180+oscillation/ minute should be used.

For quality analysis, a minimum of one reference and one blank should be included in a batch of 25 samples. Duplicates are done on approximately 5% of samples. Reproducibility of P should be with in ± 10%.

Bray and Kurtz P_1 Method

The dilute acid-fluoride extractant removes easily acid soluble P from phosphates bound to Al, Fe and Ca. Phosphate in the extracts is determined colorimetrically as phosphomolybdenum blue with ascorbic acid as a reducing agent and antimony (Sb) added to give a stable Mo-P-Sb compound.

19. Soil Potassium

13.1. Determination of Available Potassium

The determination of plant available K is complicated as, in addition to K soluble in water, it includes part of the exchangeable K. Another factor which complicates the determination of plant available K in arid and semi-arid regions is that extracting solutions used for exchangeable K may also extract part of the non-exchangeable K from K-aluminium silicate clay minerals, like K-feldspars and micas. Plants in the field do not benefit from this form of K. Moreover, moisture content of the soil sample, and the manner of drying the sample also affects K extractability. In general, routine laboratory tests for determining plant available K do not reflect the true situation under field conditions, because of the variation in the clay mineralogy of the soils. The following procedure is suggested for determining plant available K in soils of the arid and semi-arid regions, and it may also apply to Ca and Mg.

Apparatus

- Centrifuge
- Round bottom centrifuge tubes - 50 ml
- Mechanical shaker
- Flame photometer

Add here

Reagents

1. Ammonium acetate (NH4OAc), 1.0 M solution: Add 57 ml of conc. Acetic acid to 700-800 ml distilled water in a 1-litre beaker. Then, with constant stirring, add 68 ml of conc. ammonium hydroxide. Bring the volume close to 1 litre, and adjust pH to 7.0 by the addition of more ammonium hydroxide or acetic acid. or, dissolve 77.09 g of ammonium acetate in distilled water and dilute to 1 litre. Adjust pH to 7.0.
2. Potassium chloride (KCl) 0.02 M in 1.0 M ammonium acetate: Dissolve 1.491 g of potassium chloride in one litre of 1.0 M NH4OAc solution to be used for preparing standard curve.
3. Lithium Chloride (LiCl), 0.05 M: Dissolve 2.12 g of lithium

chloride in distilled water and dilute to 1 litre.

Procedure

1. Weigh accurately 5 g of soil and transfer into a 50 ml centrifuge tube.
2. Add 20 ml of 1.0 M ammonium acetate solution to the tube; stopper and shake in a reciprocal shaker for 5 minutes.
3. Centrifuge at 2000 rpm for 5 minutes or until the supernatant is clear.
4. Decant the supernatant into a 100 ml volumetric flask.
5. Repeat steps 2- three more times.
6. Make up the supernatant volume to 100 ml by adding ammonium acetate solution.
7. Prepare a series of working K standard solutions in the range of 0-2 meq/l of K from stock solution of 0.02 M KCl already prepared. For better results, add LiCl in each standard to yield a final concentration of about 5 meq/l of LiCl.
8. Determine K concentration in the extract by flame photometer.

Calculations

Meq of K/100 gm soil = reading in meq/litre X ml of diluted

$$\text{Solution} \times \frac{\text{Total volume of extract}}{\text{Stock solution taken}}$$

20. Soil Sulphur

Sulphur (S) is present in soils in organic and inorganic forms. Organic S is an important constituent of proteins and amino acids. The major inorganic sources of S include gypsum ($CaSO_4$), and pyrite (Fe_2S). Sulphur is also added to soil as a result of the use of fertilizers containing S, such as K_2SO_4, and some pesticides. It exists in soil and soil solution mainly as the SO_4^{2-} anion in combination with the cations, Ca^{2+}, Mg^{2+}, K^+, Na^+ or NH_4^+. When present in the form of elemental sulphur (S), it will be oxidized under aerobic condition to formSO_4^{2-}. Under anaerobic condition, SO_4^{2-} will be reduced by microorganisms into SO_3^{2-}-and S_2^-. Hydrogen sulphide (H_2S) is formed especially in swamps and other areas with stagnant water. Water-logged and paddy soils provide a suitable environment for the formation of H_2S. In soils rich in iron, H_2S usually precipitates as FeS, which imparts to soils a black colour. When H_2S is allowed to accumulate, it is not only toxic to soil organisms, but also it creates environmental problems. A group of bacteria present in soils is capable of oxidizing the H_2S into elemental S and SO_4^{2-}. Away from arid and semi-arid regions, probably in some acid soils, further formation of sulphuric acid (H_2SO_4) will lower soil pH to about 2, resulting in the formation of acid sulphate soils, sometimes called cat-clays. In addition to the toxicity created by the extremely low pH, the acidity will liberate very high levels of Al and Fe, which may reach toxic levels to plant growth. The total S content in soils varies widely from soil to soil. Sandy soils in the humid regions are generally low in S (0.002%). In contrast, soils in arid regions may contain 5% SO_4^{2-}-S. Plants absorb S mostly in the sulphate SO_4^{2-} form, which is the available form of S. Sulphate is extractable by water, NaCl,$CaCl_2$, NH_4OAC, $NaHCO_3$ and $Ca(H_2PO_4)^2$ solutions. Concentration of SO_4^{2-} is determined as follows:

Principle

Besides some amount in the soil solution, available sulphur in mineral soils occurs mainly as adsorbed $SO_4^=$ ions. Phosphate ions (as monocalcium phosphate) are generally preferred for replacement of the adsorbed $SO_4^=$ ions. The extraction is also carried out using $CaCl_2$ solution. However, the former is considered to be better for more efficient replacement of $SO_4^=$ ions. Use of Ca salts have a distinct advantage over and leads to easy filtration $SO_4^=$ in the extract can be estimated turbid metrically using a colorimeter/spectrophotometer.

A major problem arises, when the amount of extracted sulphur is too low to be measured precisely; To overcome this problem, addition of seed solution of known S concentration is added to the extract to raise concentration to easily detectable level. Sulphur in the extract can also be estimated by a colorimetric method using barium chromate (Nemeth, 1964; Palaskar *et al.*, 1981), but the turbidimetric method (Chesnin and Yien, 1950) given below is mainly used in the soil testing laboratories.

Instruments

i) Colorimeter or spectrophotometer or autoanalyzer.

ii) Mechanical shaker.

Reagents

1. **Mono-calcium phosphate extracting solution (500 mg P L^{-1}):** Dissolve 2.035 g of $Ca(H_2PO_4)_2.H_2O$ per litre.
2. **Gum acacia-acetic acid solution:** Dissolve 5 g of chemically pure gum acacia powder in 500 mL of hot water and filtered in hot condition through Whatman No. 42 filter paper. Cool and dilute to one litre with dilute acetic acid.
3. **Barium chloride:** Pass AR grade $BaCl_2$ salt through 1 mm sieve and store for use.
4. **Standard stock solution (2000 mg S L^{-1}):** Dissolve 1.089 g of oven dried AR grade potassium sulphate per 100 mL.
5. **Working standard solution (10 mg S L^{-1})**: Measure exactly 2.5 mL of the stock solution and dilute to 500 mL.
6. **Barium sulphate seed suspension:** Dissolve 18 g of AR grade $BaCl_2$ in 44 mL of hot water and add 0.5 mL of the standard stock solution (given above). Heat the contents to boiling and then cool quickly. Add 4 mL of gum acacia-acetic acid solution to it. Prepare a fresh seed suspension for each estimation everyday.
7. **Dilute nitric acid (approx 25%):** Dilute 250 mL of AR grade conc. HNO_3 to one litre.
8. **Acetic-phosphoric acid:** Mix 900 mL of AR grade glacial acetic acid with 300 mL of H_3PO_4 (AR grade).

Procedure

1. Weight 20 g of soil sample in a 250 mL conical flask.
2. Add 100 mL of the monocalcium phosphate extracting solution (500 mg P L^{-1}) and shake for one hour. Filter through Whatman No. 42 filter Paper.
3. Measure 10 mL of the clear filtrate into a 25 mL volumetric flask.
4. Add 2.5 mL of 25% HNO_3 and 2 mL of acetic-phosphoric acid. Dilute to about 22 mL, stopper the flask and shake well.
5. Shake the $BaSO_4$ seed suspension and then add 0.5 mL of it and

0.2 g of $BaCl_2$ crystals. Stopper the flask and invert three times and keep.

6. After 10 minutes, invert 10 times and keep. After another 5 minutes, invert 5 times.
7. Allow to stand for 15 minutes and then add 1 mL of gum acacia-acetic acid solution.
8. Make up the volume, invert three times and keep aside for 90 minutes.
9. Invert 10 times and measure the colour intensity at 440 nm (blue filter).
10. Run a blank side by side.

Preparation of standard curve for S

1. Place 2.5, 5.0, 7.5, 10.0, 12.5 and 15.0 mL portions of the working standard solution (10 mg S L^{-1}) into a series of 25 mL volumetric flasks to obtain 25, 50, 75, 100, 125 and 150 µg S.
2. Proceed to develop turbidity as described above for sample aliquots.
3. Read the colour intensity and prepare the curve by plotting readings against sulphur concentration (In µg in the final volume of 25 mL)

Calculation : Available S in soil (mg kg^{-1}) $= \frac{R \times 10}{10 \times 20}$

Where, r stands for the quantity of S in mg as obtained on X-axis against a reading.

❑❑❑

Chapter 13

Soil Micronutrients

13.1. Iron, Zinc, Manganese and Copper

13.1.1. Iron

Iron (Fe) is the third most abundant element in earth crust. The most important iron-minerals in soil are *hematite* (Fe_2O_3) and *magnetite* (Fe_3O_4). The hydrated form of hematite is often called *limonite* ($2Fe_2O_3.3H_2O$). Haematite is red in colour and its presence gives the soils a red colour. Magnetite is black in colour and crystalline in nature. It has strong magnetic properties. Another iron mineral is *pyrite* (FeS_2), which occurs in soils as yellow crystals with a metallic luster similar to gold, hence the name *fool's gold. Ilmenite* ($FeTiO_3$) is another important iron mineral, and is a valuable source for the production of many industrial products. In addition, Iron occurs in soils in an amorphous form coating soil particles and, also, as organic complexes. The granular soil structure in Oxisols is believed to be stabilized by iron coatings around the structural units. In soil solution iron exists in two oxidation states, ferrous Fe^{2+} and ferric Fe^{3+} Under anaerobic conditions, Fe^{2+} is the dominant species, whereas under aerobic conditions Fe^{3+} is the dominant species. The concentration of Fe^{2+} and/ or Fe^{3+} in soil solution is very low, due to the highly insoluble nature of Fe_2O_3 minerals. The concentration of soluble Fe in river water and ground water is estimated to range from 0.1 to 10 mg/l (Krauskoph, 1973; Tan, 1994).

Iron is an essential micronutrient for plant growth. It is needed for chlorophyll formation. In animals, iron is present in the blood haemoglobin, which acts as a carrier of oxygen. The normal iron concentration in plant tissue varies considerably with plant species from 100 μg/g dry matter of grasses to 1000 μg/g in alfalfa dry matter. Iron concentration of <20 μg/g in plant tissue is considered deficient, and may result in development of chlorosis, frequently manifested as yellow stripes on young leaves. Iron deficiency in soils can be induced by over liming, and the presence of high amounts of phosphates. Iron deficiency often occurs on plants growing in alkaline calcareous soils. Therefore, iron deficiency is most likely a problem in the calcareous soils of arid and semi-arid regions.

13.1.2. Zinc

Zinc (Zn) is present in small quantities in igneous and sedimentary rocks. Concentration of Zn is usually higher in basic igneous rocks, such as basalt, than in acid igneous rocks, such as granite. The major inorganic sources of Zn are *sphalerite*, ZnS, *smithsonite*, $ZnCO_3$, and *hemimorphite* $[Zn_4(Si_2O_7)(OH)_2.H_2O]$. *Hydrozincite* $[Zn_5(CO_3)_2(OH)_6]$ is another Zn mineral frequently mentioned, but this mineral is considered similar to smithsonite (Hurlbut and Klein, 1977). All the minerals stated above are very rare in soils, and occur only in large amounts as Zn ore deposits.

The concentration of total Zn in soils is approximately in the range of 10 μg to 300 μg/g. The average Zn content in normal agricultural soils is 50 μg/g (Brady, 1990). In soil solution, Zn is present as the divalent Zn_{2+} ion. Most Zn compounds are sparingly soluble in water. Because of its cationic nature, the Zn ion is adsorbed by the negatively charged surfaces of soil colloids, and is then called *exchangeable Zn*. The concentration of Zn ions is dependent upon pH, as is the case with all the other microelements. At low pH, Zn^{2+} concentration is very high in soils, whereas at high pH, Zn^{2+} concentration is low. At pH 9.5, Zn^{2+} exhibits minimum solubility and precipitates as $Zn(OH)_2$. Therefore, high soluble Zn contents occur in acid soils, whereas low soluble Zn contents are found more in basic soils, especially in Aridisols. For example, the concentration of Zn^{2+} ions in neutral and alkaline calcareous soils is in the range of 1 u 10-8 to 1 u 10-10 *M* (Norvell, 1973). Consequently, Zn deficiency is found in alkaline calcareous soils. Zinc is a micronutrient to plants, and is needed only in very small amounts. The element functions as a catalyst. It is present in several plantenzymes, e.g. dehydrogenase, proteinase, and peptidase. Zinc is also essential for seed and grain production, and development of growth hormones.The normal concentration of Zn in most plants ranges on the average from 15 to 125 μg/g. Zn concentration of <15 μg/g of leaf dry matter cause Zn deficiency, which is manifested as stunted growth. On maize plants, Zn deficiency produces white to yellow leaves with bleached stripes, a symptom known as *white bud* of maize. On the other hand, Zn concentration of > 400 μg/g of leaf dry matter is considered excessive and may induce

Zn toxicity. Zinc deficiency is common in the calcareous soils of arid and semi-arid regions.

13.1.3. Manganese

Manganese (Mn) is present in small quantities in many rocks. It is released into the soil by rock weathering and is re-deposited in various forms of Mn oxides. The total Mn content in soils varies considerably from 20 μg/g to 6000 μg/g (Krauskoph, 1973). The element can exist in three oxidation states: Mn^{2+} Mn^{3+}, and Mn^{4+}. The divalent manganese ion is the main form of Mn in soil solution, especially in reduced soil environment. Because of its cationic nature, Mn^{2+} is usually adsorbed on the negatively charged surfaces of soil colloids. It is then called *exchangeable manganese.* The trivalent form usually exists as Mn_2O_3, which can be found in substantial amounts in acid soils. The trivalent ion itself is unstable in soil solution. The tetravalent form (MnO_2), is perhaps the most stable and inert form of manganese. The concentration of Mn^{2+} in soil solution is very low, seldom exceeding 0.05 mg/l. Its concentration increases at low pH values, and decreases at high pH values. Therefore, high amounts of manganese may be present in highly weathered acid soils, such as Ultisols and Oxisols. On the other hand, low amounts of manganese usually occur in Aridisols. Manganese is an essential micronutrient and is needed by plants to activate a number of enzymes. It also plays an essential role in photochemistry and in N-metabolism and assimilation. Manganese is required for plant growth only in very small amounts. The normal Mn content in plant tissue is 20-500 μg/g dry matter. A Mn concentration of < 20 μg/g dry matter is considered deficient and is usually manifested by development of necrotic spots, known as *marsh spots* in pea leaves, and *grey specks* in oat leaves. Soyabean and oats are especially sensitive to Mn deficiency. Manganese deficiency is likely to occur in calcareous soils of arid and semi-arid regions.

13.1.4. Copper

Copper (Cu), is classified as a *native element,* meaning that it can occur as a native element in the earth crust in contrast to such elements as Al, which exist only in the form of a compound. The most common form of Cu in soils is in the form of minerals, e.g. in the forms of sulphides, sulphates, oxides, carbonates and silicates. The most common Cu mineral in soils is perhaps chalcopyrite ($CuFeS_2$). The total Cu content in soils is in the range of 10 μg/g to 80 μg/g

(Krauskoph, 1973). In the soil solution, copper exists as Cu^{+} (cuprous), or Cu^{2+} (cupric) ions. At concentrations > 10-7 mol/l Cu^{+} ions are unstable at ordinary temperature. However, it can exist at these conditions as a $CuCl_2$- chelate (Krauskoph, 1973). Cuprous ions are so unstable in aqueous solution that they will be automatically affected by the soil redox process, called *auto-reductionoxidation* reaction, and converted into Cu and Cu^{2+}. Therefore, cupric ions are more stable and are the major copper ions in the soil solution. However, Lindsay (1973) noted that Cu^{2+} is the dominant ion in soils at pH values < 7.3, whereas above this pH, Cu (OH)+ is the major copper ion. The concentration of soluble Cu ions in soils is on the average 20 mg/l. Because of its cationic nature, Cu^{2+} will be adsorbed by negatively charged surfaces of soil colloids, and are then called exchangeable Cu. The concentration of free and exchangeable Cu ions is high in acid soils, and low in basic soils. The Cu^{2+} concentrations in many neutral and calcareous soils are reported to amount only 1 μ 10-12 M (Norvell, 1973). Therefore, Cu deficiency is more likely to occur in basic soils, e.g. Aridisols,whereas Cu toxicity may be displayed more by highly weathered acid soils, e.g. Ultisols and Oxisols. Copper is a micronutrient, and is needed only in small amounts by plants. The element is required for chlorophyll formation, hence Cu affects photosynthesis. It is essential in enzyme reactions, and in the reproductive stages. Cu is noted to be needed in protein and carbohydrate metabolism, and in nitrogen fixation. The normal concentration in plant tissue is approximately 4-30 μg Cu/g. Present at concentration < 3 μg/g dry matter, Cu deficiency will occur, which is manifested by yellowing and curling of leaves, stunted growth and development of short internodes. On the other hand, Cu concentration in leaf tissue of >20 μg/g may result in Cu toxicity, which is manifested by yellowing of leaves, and poorly developed roots with frequently discoloration.

13.2. Determination of Available Micronutrients

Available Fe, Zn, Cu and Mn in soils are extracted with several chelating agents, including EDTA, EDDHA and DTPA. Lindsay and Norvell (1978) developed the DTPA (diethylenetriaminepentaacetic acid) method and reported good correlations between DTPA-extractable micronutrients and plant growth. These elements can also be determined in the soil solution extracted by centrifugation of a moist soil sample collected directly from the field.

Atomic absorption spectrophotometer: Atomic absorption spectrophotometer is based on the principle that when atomic vapours of an element are irradiated by the radiation of a characteristic wavelength (i.e. the light from a source whose emission lines are those of the element in question), they absorb in direct proportion to the concentration of the element being determined.

Instrument features : A wide range of atomic absorption spectrophotometer is available today, all of them have the basic features in common and consist of the following components :

a) **A Light source :** A Light source emits the spectrum of the element to be determined. The most widely used light source is hollow cathode lamp which is designed and operated in such a way that the lines to be measured are sharp, of stable intensity and free from background.

b) **Atomizer-Burner assembly :** A means of producing atomic vapours of the element to be analyzed. The solution to be analyzed is drawn by capillary and converted into stream of compressed air to a fine spray which after condensation of larger droplets is mixed with the fuel gas acetylene and burnt in a long flame (at 2100-2400°C) in a stainless steel burner.

c) **A Monochromator :** It isolates the absorbing resonance lines from other non-absorbing lines. When the light coming from the HCL, after traversing the flam, enters the monochromator which is already set at the wavelength of the resonance lines of the desired element, the monochromator performs its function.

d) **A Detector :** It measures the magnitude of absorption of the characteristic radiation.

e) **A Photomultiplier Tube :** It amplifies the absorption signal and converts the light radiation into electrical energy.

f) **A readout system :** It measures the absorbance in volts. It is normally a strip chart recorder, a digital display, a meter or printer. The presently available AAS have features like automatic calibration with one or more standards, automatic curve corrections, automatic and foolproof gas switching and calculation of average and standard deviations in repetitive runs. switching and calculation of average and standard deviations in repetitive runs.

Collection and preparation of soil and plant samples : To avoid contamination, soil samples are to be collected in plastic tub, using rust free instrument or wood and kept in polythene lined cloth bags. Samples are prepared with the help of wooden mortar and pastle and sieved through 2mm nylon screen/mosquito net cloth or stainless steel sieve.

Similarly, plant samples (leaves, grains or straw) should be washed with 0.01N HCl, rinsed with glass distilled water dried in oven at 65°C and crushed with the help of stainless steel scissors.

Soil extraction : DTPA offers the most favourable combination of stability constants for the simultaneous complexing of Zn, Cu, Fe and Mn, Cd, Co, Ni and Pb (Lindsay and Norvell, 1978). Buffering of extractant in a slightly alkaline pH range (7.3) by including soluble Ca^{2+}, avoids the dissolution of $CaCO_3$ with the release of occluded micronutrients due to CO_2 partial pressure of approximately 10 times that in atmosphere, as the soil contains slightly higher CO_2 levels than found in the atmosphere.

a) **Extracting solution** : (0.005 M DTPA) Dissolve 1.9679g of DTPA (Diethylene tri amine penta acetic acid) + 13.3 ml TEA (Triethanol amine) + 1.47g $CaCl_2 .2H_2O$ in 200 ml distilled water, dilute to 900 ml, adjust pH 7.3 with 6N HCl while stirring and then make upto 1 liter and mix thoroughly.

b) **Apparatus required** : Shaker (Horizontal or Rotatory), iodine value flasks (100 ml capacity) or conical flasks with glass stoppers, funnels, filter paper whatman No.1, plastic storage bottles and Atomic absorption spectrophotometer.

Stock Standard Solutions : The standard solutions of different micro-nutrients should preferably be prepared by using their wires. Dissolve 1g wire in a minimum volume of 1:1 nitric acid and dilute to 1000ml with distilled water to obtain 1000 μg/ml solution of micro-nutrient, or take salts of metals as follows:

Zn- 4.398g l^{-1} $ZnSO_4,7H_2O$

Cu- 3.929g l^{-1} $CuSO_4,5H_2O$

Fe- 4.977g l^{-1} $FeSO_4,7H_2O$

Mn- 3.598g l^{-1} $MnSO_4,H_2O$.

The prepared standards are also available in the market. Out of these standards, prepare working solution of 50 ppm. Then a series of standard solution of 0.5, 1.0, 1.5, 2.0 and 2.5 ppm may be prepared for each metal.

d) **Background correction** : The reading of a spectral line always includes any contribution from the flame and sample matrix. Failure to correct properly for the background reading can be a source of serious error. Although the need for fast background correction is most obvious with graphite furnace work, it is also a consideration with flame atomic absorption.

The most common method of background correction in atomic absorption spectrometry involves the use of a continuum source, such as a deuterium lamp to measure the background. The source used is a deuterium filled discharge lamp, which emits an intense continuum spectrum from 190 nm to about 400 nm. This is the region where most atomic absorption lines occur and where the effects of background absorption are most pronounced. The poly-atomic gas D_2, is used in the lamp because a continuum is produced rather than a line spectrum.

The deuterium lamp is different from a hollow cathode lamp in construction and operation. The lamp incorporates a heated, electron-emitting cathode, a metal anode and a restrictive aperture between the two. A discharge current of several hundred milli amperes excites the deuterium gas. The discharge is forced to pass through the small aperture, forming a defined area of high excitation and hence, high light emission. A suitable window transmits the light to the spectrometer's optical system.

To obtain successful background correction the deuterium lamp must be correctly aligned, and its intensity must be matched to that of the hollow cathode lamp. It is important that both the deuterium source and the hollow cathode source are aligned to follow the same optical path. If they are not, then the two measurements may not be made on the same atom population and significant errors may occur.

In order to balance the intensity of the deuterium lamp with the hollow cathode lamp, it may be necessary to change the hollow cathode lamp current to a higher or lower value depending on the relative intensities of the lamps. Although most modern AA spectrophotometers incorporate so called "simultaneous" background correction, they rely on two measurements separated slight in time. One measurement is of the total absorbance (atomic plus background) and the other is of the background only. The background is electronically subtracted from the absorbance to give the background corrected atomic absorbance with the continuum source method of background correction, the total absorbance is measured during the hollow cathode lamp pulse and the background during the deuterium lamp pulse. With the Zeeman method using a modulated magnetic field, the total absorbance is measured with the magnetic field off and the background with the field on.

e) **Soil analysis** : Weigh 12.5g soil sample in 100 ml iodine value flasks. Add 25 ml DTPA solution. Shake this mixture for 2 hours on shaker at 70 to 80 oscillation per minute, filter through acid washed distilled water rinsed, whatman No.1 filter paper and collect the filtrate in plastic bottles. Determine the content of micronutrients on atomic absorption spectrophotometer.

Factors : For soil multiply the concentration read on AAS computer sheet by "2".

OR

Procedure

1. Weigh 5 g of air-dried soil (<2 mm) in a 100 ml polyethylene centrifuge tube, add 20 ml DTPA solution, and shake for 30 minutes on a mechanical shaker.
2. Centrifuge, and decant into a sample bottle fitted with funnel and filter paper.
3. If needed, dilute the extract so that the reading is in the linear working range of the atomic absorption spectrophotometer.

Estimation of Boron in Soils

Boron occurs as anion in soils and is required by plants in very small quantity. Water soluble B makes the estimate of its availability to plants. Total boron in soils varies from 20 to 200 mg kg^{-1} and available (water soluble) boron in soils ranges from 0.03 to 12 mg kg^{-1} respectively. The threshhold value ranging from 0.1 to 0.5 mg kg^{-1} (water soluble B) depends upon the soil type, crops, and other factors, below which the response to applied boron may be expected. Some sensitive crops to boron deficiency are listed in table 1. Its availability is affected by soil pH as under:

- Deficiency of B is generally observed in old acid leached soils.

- Availability increased with the rise in soil pH having significant positive correlation with pH rising from 4.7 to 6.7.
- In neutral, saline and calcareous soils the B availability again decreases with the rise in soil pH having significant negative correlation with the rise in pH from 7.1 to 8.1. In calcareous soils B fixation occurs with the condensation of borate radical into long chains in the presence of Ca.
- In alkaline soils the availability of B is high and may be even toxic for plant growth.

Besides this the low moisture availability also causes B deficiency. Irrigation water containing Boron between 0.3 to 0.6 mg kg^{-1} can be used safely, whereas, irrigating soils with water containing 1 to 3 mg kg^{-1} B causes toxicity of B in plants.

Boron determination (Azomethine H Method)

Azomethine H forms coloured complex with H_3BO_3 in aqueous media. Over a concentration range of 0.5 to 10 μg B/ml the complex is stable at pH 5.1. Maximum absorbance occur at 420 nm with little or no interference from a wide variety of salts. This technique is rapid, reliable and more convenient to use than traditional procedures employing carmin, curcumin or quinalizarin (John *et al.*, 1975).

Apparatus

1. Spectrophotometer
2. Poly-propylene tubes 10 ml capacity.

Reagents

1. Distilled water
2. Buffer solution : Dissolve 250 g of ammonium acetate (NH_4OAc) and 15 g of ethylenediaminetetracetic acid (EDTA disodium salt) in 400 ml of distilled water. Slowly add 125 ml of glacial acetic acid and mix.
3. Azomethine H reagent : Dissolve 0.45 g of azomethine H in 100 ml of 1% L ascorbic acid solution. Fresh reagent should be prepared weekly and stored in a refrigerator.
4. Calcium hydroxide suspension : Add 0.4g $Ca(OH)_2$ to 100 ml distilled water.

5. 0.1 N HCl : Add 8.3 ml conc. HCl to 900 ml distilled water, mix, cool to room temperature and make up the volume to 1000 ml.
6. Calcium chloride 0.01 M Dissolve 1.11 g of anhydrous $CaCl_2$ in 900 ml distilled water and make up the volume to 1000 ml.
7. Boron standard solution : Dissolve 0.114 g of Boric acid (H_3BO_3) in distilled water and adjust the volume to 1000 ml. Each ml contains 20 µg B. Dilute 10, 20, 30, 40 and 50 ml of the stock solution to 100 ml with distilled water to have solution with B concentration of 2, 4, 6, 8 and 10 µg of B/ml, respectively. Include a distilled water sample for the 0.0 µg of B/ml standard solution.

Procedure

Take 1 ml of aliquot of blank and diluted B standards into a 10 ml polypropylene tube, add 2 ml of buffer solution and mix. Add 2 ml of azomethine H reagent, mix and after 30 minutes read the absorbance at 420 nm on spetrophotometer. With the help of absorbance readings of standard solutions of different concentration of B the standard curve is drawn and a factor for concentration of B for 1 absorbance is calculated, which is utilized to calculate B in the soils, plant or water sample.

Preparation of Extracts

1. **Soil extracts** : The hot water soluble extraction procedure of Berger and Truog (1939) is being used widely with slight modification of adding dilute electrolyte (0.01 M $CaCl_2$) instead of water only. This provides clear, colourless extract which eliminates the need of charcoal for decolourzation. Beside this a negative error, associated with B adsorption by charcoal, is also removed. Place 20 g air dry soil in 250 ml low B flat bottom flasks and add 40 ml of 0. 01 M $CaCl_2$ solution. Attach water cooled reflux condenser to the flask. Heat the flasks for 5 minutes and then cool and filter the suspension in plastic bottles.
2. Transfer 20 ml aliquot to evaporating dish, add 2 ml $Ca(OH)_2)$ suspension and evaporate the solution to dryness. Heat the evaporating dishes gently to destroy organic matter, cool to room temperature, add 5 ml 0.1N HCl. Triturate the residue with rubber policeman to ensure the complete dissolution of the residue (Bingham, 1982).

3. For analysis of B pipette 1 ml of the aliquot and proceed as for the standard curve.

22.2.2. Measurement of Element Concentration Using Atomic Absorption Spectroscopy

Atomic absorption spectrophotometry has become a common practice in almost all laboratories, especially for the measurement of trace elements (Te) concentration in solution. Each instrument, AA spectrophotometer, has its instruction manual that guide the user to the adjustment and operation of the instrument, but it is essential to have a general knowledge of the basic principles of the technology. Atomic absorption spectroscopy (AAS) uses absorption of light to measure the concentration of analyte atoms in a flame or graphite furnace. The light source is usually a hollow-cathode lamp of the element that is being measured. Lamps convert electrical energy into radiation. Atoms absorb the radiation and make transitions to higher energy levels. Light absorption is proportional to the amount of analyte atoms in the path of light. Concentration measurements are determined from a working curve after calibrating the instrument with standards of known concentration. Atomic absorption spectroscopy requires that the analyte atoms be in the gas phase. Ions or atoms in a sample must undergo vaporization or atomization in a high-temperature source, such as a flame or graphite furnace. Flame AAS uses a slot type burner to increase the path length, and therefore to increase the total absorbance. Sample solutions are usually aspirated with the gas flow into a nebulizing/mixing chamber to form small droplets before entering the flame. The furnace AAS is a much more efficient atomizer than the flame and can directly accept very small quantities of sample. Samples are placed directly in the graphite tube and the furnace is electrically heated in several steps to dry the sample, ash organic matter, and vaporize the analyte atoms. While flame AAS measures concentration of analyte in μg/ml, furnace AAS detects concentrations in μg/l.

A calibration curve is a plot of the analytical signal (the instrument or detector response) as a function of analyte concentration. These calibration curves are obtained by measuring the signal from a series of standards of known concentration. The calibration curves are then used to determine the concentration of an unknown sample, or to calibrate the linearity of an analytical instrument.

Procedure

1. Prepare an intermediate standard solution by pipetting 10 ml from the 1000 μg/ml Stock solution of the analyte Te into a 200 ml volumetric flask, and dilute to the volume with DTPA solution.
2. Prepare standard solutions in the working range, like 0, 1, 2, 5, 10 μg/ml of the trace metal. Always dilute with the DTPA solution.
3. Now follow the step by step procedure given in the instruction manual to optimize the working condition of the instrument.
4. Measure the signals from the series of working standards of known concentration, and plot the analytical signals (the instrument or detector response) as a function of analyte concentration.

Calculations

5 g soil

g Te/g soil g Te/ml X 20ml sample.

❑❑❑

Chapter 14

Water Analysis

Water quality assessment for irrigation

Irrigation water quality refers to its suitable for use. A good water quality has the potential to allow maximum yield under good soil and water management practices. However, with poor quality water, problems of salinized soil impaired cropping and reduced yields can be expected to develop unless special management practices are adopted to maintain or re-store maximum soil production and production capability under a given set of conditions. Therefore, the quality of irrigation water is an important factor for sustenance of long-term soil productivity. Use of poor quality water for a long time can render the soil less productive or even barren. Presence of the higher amounts of nutrient ions low or marginally saline waters sometimes appear to stimulate crop growth. However, excess of the soluble slats in water leads to their accumulation in the surface layer particularly in fine-textured or poorly drained soils. Many areas in the country are facing a serious problem of not only scarcity of water, but also of its extremely poor quality. Tubewell or well waters generally pose such problems more than canal waters, especially in arid and semi-arid regions. It is, therefore, advisable to get the water tested for quality while installing a tube well instead of repenting later. Locating appropriate strata during boring for well is to be done by the 'hit and trial' method causing huge investments, if the knowledge about surrounding localities is lacking. In such a situation samples of water should be tested while installation is in progress and the appropriate depth decided accordingly. Irrespective of the source of irrigation water all that is required to be done is to draw a proper sample for testing in the laboratory.

Water sampling techniques

Irrigation water, irrespective of its source, always contains some soluble salts. The suitability of waters for a specific purpose depends upon the types and amounts of dissolved salts. Some of the dissolved salts or other constituents may be useful for crops but the quality or suitability of waters for irrigation purposes is assessed in terms of the presence of undesirable constituents and only in limited situations, the irrigation waters is judged as a source of plant nutrients. Some of the dissolved ions, such as NO_3 are useful for crops. The limits of purity established for drinking water, water to be used for industrial purposes and for agriculture are different. It may, therefore, be possible that a water which is not good for drinking and industrial use, may be quite suitable for irrigation. This publication deals with irrigation waters only. The most important characteristic that determine the quality of irrigation waters are:

1. pH
2. Total concentration of soluble salts judged through electrical conductivity (EC).
3. Relative proportion of sodium to other cations, such as Ca and Mg referred as Sodium Adsorption Ratio (SAR).
4. Concentration of boron or other elements that may be toxic to plants
5. Concentration of carbonates and bi-carbonate as related to the concentration of calcium plus magnesium referred as Residual Sodium Carbonate (RSC).
6. Content of anions, such as chloride, sulphate and nitrate.

The analytical data on the above parameters are used to describe the quality of irrigation water taking standards fixed for each aspect as an index. Some waters coming from industries as effluents and domestic waste water as sewage may contain some specific plant nutrients also. These waters may be useful for irrigation of field crops, if assessed for toxic/pollutant metals, and organic and microbial constituents with regard to their suitability or otherwise. Determination of organic constituents is generally carried out under two categories: i) organic substances which quantify an aggregate amount of organic carbon, and ii) individual or specific organic substances, such as benzene, DDT, methane, phenol, endosulfan, etc. Important determinations are the chemical oxygen demand (COD) which gives the total organic substances and the bio-chemical oxygen demand (BOD), which gives the amount of total biodegradable organic substances in the water sample. To keep the scope of this publication tailored to Service laboratories, only parameters of practical utility to judge the quality of commonly used irrigation waters are described. Thus, the aspects pertaining to the use of effluents and sewage waters have not been covered. Similarly, the aspect of assessing water as a carrier of plant nutrients also has not been covered.

a) Important indices of practical utility to judge irrigation water quality

The following standards are laid down. Run the tubewell/hand pump for about 15-20 minutes to drain out the water retained in the pipe. Wash the bottle repeatedly with the water before taking about 500mL of sample. The bottle should not be washed with detergent or soap :

1. If a new tube installation in progress, collect water samples at different depths at intervals of about 3-4 metres. Since the water in such a situation is always turbid with suspended impurities, the samples may be collected as such without caring for turbidity, as it does not affect the test results. It is to be filtered before analysis any way.
2. If water from a tank or pond is to be tested, collect the sample from at least 5 to 10 meters away from the boundaries. For this purpose tie a suitable (small) container at one end of a long bamboo pole and collect the sample after displacing surface water, which might contain organic material floating over it.
3. From a well, the sample can be drawn either during irrigation, just before the water falls in the channel or by drawing it with the help surface should be disturbed a little to remove any floating material before collection of the sample.
4. Place the cap on the bottle tightly.
5. Take a piece of paper of suitable size (about 8 cm x 4 cm), write the name, address, sample identification mark etc. and paste it firmly on the bottle.
6. Write separately on a sheet of paper the name of the crop to be grown, texture of soil to be irrigated, previous experience about the effect of the water on soil surface, if any, earlier test results, if available, other sources available for irrigation, and crop

performance so far. Send it to the laboratory along with the sample, preferably within 2-3 days of collection of the sample.

7. In case any delay is anticipated in sending the sample for testing or the samples are to be stored for more than a week, add 2 or 3 drops of toluene to the sample to prevent microbial growth or preferably draw a fresh sample.

8. It is desirable to send both soil and water samples together for testing, as it helps in making rational recommendations.

b) Chemical analysis of water

1. Soil reaction (pH)

The determination of pH of water samples is based on the same principle as given for soil section 1.2.2.

Instrument

1. pH meter
2. **Reagents :** Buffer solutions of pH 4.0, 7.0 and 9.2

Procedure

1. Standardize the pH meter with the three pH buffer solutions after giving warm up time, if needed.
2. Take about 50 mL of water sample in a 100 ml beaker (preferably plastic), immerse the electrodes of the pH meter and record the pH value.

2. Soluble Salts

Principle

The total soluble salts can be determined as the electrical conductivity (EC) of the water sample using a conductivity meter/ solubridge as in case of soil-water suspension. The electrical conductivity is directly proportional to the amount of salts (ions) and is expressed as dS m^{-1} (which is equivalent to the unit cm^{-1} (. Usually, a temperature of 20 °C is kept for the EC determinations.

Instrument

1. Conductivity Bridge (EC meter)

Reagent

1. Switch-on the conductivity meter and allow warm up time, if required.
2. Standardize the EC meter with 0.01 *N* KCL which gives a value of 1.41 dS m^{-1} at 25°C, if the cell constant is 1.0.
3. Fill the conductivity cell or dip the cell assembly in water sample taken in a 50 or 100 mL beaker and record the conductivity. If the value is too low, change the position of the knob suitably.

Alternatively, evaporate a known volume of water and weigh the dry residue (gravimetric method). However, the conductivity method is more precise and rapid, especially useful, when the quantity of water sample is limited and the same is to be used for further analysis.

Calculations

The EC value can either be used as such for categorizing the water on salinity basis or may be used to get the concentration as given below:

Total salt content (mg L^{-1})= EC (dS m^{-1}) at 25°C x 640

Or, Total salt content in me L^{-1} (approx) = EC (dS m^{-1}) at 25° C x 10

3. Calcium and Magnesium

Principle

The extent of sodium hazard in irrigation water is gauged in terms of the sodium concentration in relation to Ca^{2+} and Mg^{2+}. For the determination of both of them together, the versenate titration method is most popular in which EDTA- disodium salt solution is used to chelate them. The two cations can also be precisely estimated using AAS, but for all practical purposes the versenate method is good enough. Calcium alone can also be estimated by versenate method using ammonium purpurate (murexide) indicator and thus, Mg^{2+} can be obtained by subtracting Ca^{2+} from Ca^{2+} + Mg^{2+} content. Calcium estimation can also be done on a flame photometer but the precision is not very high. A number of cations including Ca^{2+} and Mg^{2+} are known to form stable complexes with versenate at different pH values. The formation of Ca and Mg complexes at pH 10 is achieved by using ammonium hydroxide buffer. A number of polyvalent ions are preferentially complexed by versenate, as these are less dissociated than those of Ca and Mg and thus, included in the titration. Fortunately the concentration of such interfering metals *e.g.* Fe, Cu, Pb, Zn, Co, Mn is quite low and negligible in most waters and can be ignored.

However, the interference, if high, can be prevented by using 2% solution of NaCN.

Reagents

1. **Standard $CaCl_2$ solution (0.01 *N*)** : As given under section 6.3.1.
2. **Standard versenate solution (0.01 *N*)** : As given under section 6.3.1.
3. **Ammonium hydroxide-ammonium chloride buffer (pH 10) :** As given under section 6.3.1.
4. **Eriochrome black T (EBT) indicator** : As given under section 6.3.1.

Procedure

1. Transfer 5 mL of the water to a 100 mL porcelain dish and about 20 mL of distilled water.
2. Add 1 mL of ammonium hydroxide-ammonium chloride buffer and 3 to 4 drops of EBT indicator.
3. Take versenate solution in a burette and titrate the contents in the dish. The wine red colour changes to light/sky blue at the end point.
4. Swirl the contents in the dish to ensure complete disappearance of the red colour.

Calculation

Ca + Mg concentration in water (meL^{-1}) = 2V

Where, V is the value of 0.01 *N* EDTA solution used (titre value) for 5 mL of water sample.

The factor '2' is arrived at from the expression given on the next page.

$N_1 V_1 = N_2 V_2$

Where, N_1 and N_2 stand for the concentrations in me L^{-1} for water and EDTA solution used, respectively, and V_1 and V_2 for the volumes of water (5 mL) and EDTA solution consumed (V).

(Normality of the EDTA solution is 0.01 *N* i.e. *10* me L^{-1})

4. Carbonates and Bicarbonates

Principle

Carbonates and bicarbonates in water can be conveniently determined by titrating a known volume of water against standard H_2SO_4 using phenolphthalein and methyl orange indicators, respectively. When a drop of phenolphthalein is added to water, the pink colour developed will indicate the presence of carbonates. In good quality water, generally pink colour will not appear, as carbonates are absent,. On titration, the pink colour disappears because all the carbonates are converted to bicarbonates as shown below:

$2NaHCO_3 + H_2SO_4 \rightarrow a_2SO_4 + 2H_2O + 2CO_2$

After completing the first titration (if pink colour appears) either the same aliquot is used for bicarbonate titration or a fresh sample is taken for it.

Reagents

1. **Standard H_2SO_4 (0.01 *N*):** Carefully add 2.8 mL of conc. Sulphuric acid t about 500 mL distilled water taken in a glass beaker and dilute on 1 l to get approximately 0.1 *N* H_2SO_4. Dilute 100 mL of this solution to 1 l. standardize it by titrating against standard alkali prepared separately.

2. **Phenolphthalein indicator:** Dissolve 0.25g of phenolphthalein powder in 100 mL of ethyl alcohol.

3. **Methyl orange Indicator:** Dissolve 0.50g of the dye in 100 mL of 95% ethyl alcohol.

Procedure

1. Transfer 5 mL of water sample to a 100 mL porcelain dish and add about 20 mL of distilled water.
2. Add 2 drops of phenolphthalein indicator.
3. If pink colour appears, titrate against 0.01 *N* H_2SO_4. Note volume of acid consumed (A mL).
4. Add 2 drops of methyl orange indicator to the colourless contents in the dish (left after the first titration or to a freshly taken 5 mL water sample). This will give yellow colour.
5. Continue the titration until the yellow colour changes to red. Notc thc volume of acid consumed in the second titration (B mL)

Calculations

Carbonates (me L^{-1})= 4A, and

Icarbonates (meL^{-1})= 2(B-A)

Where, A and B stand for the volume of 0.01 *N* H_2SO_4 (mL) consumed in the first and second titrations, respectively,

Note

The fist titre value (A) has been subtracted from the second (B), because (B) Includes the volumes consumed for already present bicarbonates as well as for those obtained from neutralization of carbonates. The basis of the factor 2 is same as described from Ca + Mg in water and it has been doubled for carbonates because only half neutralization is the first titration.

5. Sodium

Principle

Irrigation waters may have major types of hazard *viz.* salinity and sodium hazard. The latter is expressed as the residual sodium carbonate (RSC) and sodium adsorption ratio (SAR). For working out the SAR value, sodium content is determined conveniently using flame photometer or simply by subtracting the Ca+ mg content from the total soluble salts. The subtraction method, though not so precise due to the presence of other cations like K^+, is often adopted to serve the purpose. The quantity of K^+ ions inwater is generally low enough to be ignored for this purpose. The flame photometric determination of sodium is given below:

Reagents

1. **Standard stock solution (100 me Na L^{-1}):** Dissolve 5.845 g of AR grade dried NaCl in distilled wwawter and make the volume to 1 l.
2. **Working standard solution of Na :** Dilute 5, 10, 15, 20, 30, 40 and 50 mL portions of the stock solution (containing 100 me Na L^{-1}) to 100 mL in volumetric flask to get working standards of 5, 10, 15, 20, 30, 40 and 50 me Na L^{-1} concentrations.

Procedure

1. Filter a portion of the water sample, if suspended material is visible filtration is desirable, as it prevents choking of the capillary tube of the flame photometer.
2. Take the working standard solutions and record the flame photometer reading against each, after setting zero with distilled water and 100 with the highest concentration *i.e.* 50 me NaL^{-1}.
3. Feed the test sample and record the reading.
4. Draw a standard curve by plotting the reading against Na concentrations.

Calculations

Na concentration in water (me L^{-1}) is directly from X-axis against reading.

Estimation of parameters related with sodium hazards

i) Residual Sodium Carbonate (RSC)

The residual sodium carbonate may be calculated simply by subtracting the quantity of Ca + Mg from the sum total of carbonates and bicarbonates determined separately in a given sample and expressed in me L^{-1}. Thus,

$$RSC = (CO_3^{2-} + HCO_3^-) - (Ca^{2+} + Mg^{2+})$$

ii) Sodium Adsorption Ratio (SAR)

Analytical Methods

Plants are detrimentally affected, both physically and chemically, by excess salts in some soils and by high levels of exchangeable sodium in others. Soils with an accumulation of exchangeable sodium are often characterized by poor soil condition. Sodium adsorption ratio (SAR), along with pH, characterizes salt-affected soils. It is an easily measured property that gives information on the comparative concentrations of Na^+, and Mg^{2+} in soil solutions. The equation used to calculate SAR is given as follows:

$$SAR = \frac{\left[Na^{+}\right]}{\sqrt{\frac{1}{2} - ([Ca^{2+}] + [Mg^{2+}])}}$$

Where $[Na^+]$, $[Ca^{2+}]$, and $[Mg^{2+}]$ are the concentrations in mmol/ L of sodium, calcium, and magnesium ions in the soil solution. Concentrations of sodium,calcium, and magnesium are determined by first extracting the ions from the soil into solution. The solution is then analyzed to determine concentrations of the selected ions. Na^+, Ca^{2+}, and Mg^{2+} concentrations are commonly determined using atomic absorption spectrometry (AA). The SAR of a soil extract takes into consideration that the adverse effext of sodium is moderated by the presence of calcium and magnesium ions. When the SAR rises above 12 to 15, serious physical soil problems arise and plants have difficulty absorbing water.

6. Potassium

Potassium in water can easily determined by flame photometric method exactly in the same manner as sodium, except that the working standards of much lower concentrations (*e.g.* 0, 2, 4, 6, 8, 10 and 20 μK mL^{-1}) are required. These concentrations should be divided by equivalent weight of potassium (*i.e.* 39) to convert them into me L^{-1}.

Chloride

Principle

Chlorides being highly soluble are present in all waters, but the amounts are often very low in natural waters. However, their contents may be appreciable, when the electrical conductivity is high. Chlorides can be easily determined by $AgNO_3$ titration (Mohr's titration) method in which silver reacts with chloride forming AgCl precipitate. When all the chlorides are precipitated, potassium chromate (the indicator used) shows the brick red at the endpoint due to the formation of silver chromate.

Reagents

1. **Standard sodium chloride solution (0.02 *N*):** Dry AR grade NaCl in oven at 80°C for 1 hour, and dissolve 1.170 g of it in chloride-free distilled water and make to 1 l.
2. **Standard silver nitrate solution (0.02 *N*):** Dissolve 3.40 g of Ag NO_3 in about 200 mL of freshly prepared chloride-free distilled water and make to 1 l. Standardize it against 0.02 *N* NaCl solution. Store in an amber bottle away from light.
3. **Potassium chromate indicator:** Dissolve 5g of pure K_2CrO_4 in 100 mL of distilled water.

Procedure

1. Transfer 5 mL of water sample to a 100 mL porcelain dish.
2. Add about 20 mL of distilled water and 5 to 6 drops of potassium chromate indicator.
3. Titrate the contents against 0.02 *N* $AgNO_3$ solution till the brick red colour just starts appearing.

Calculations

Chloride content in me L^{-1} =4V

Where, V denotes the titre value in mL.

(1 mL of 0.02 *N* or 20 me L^{-1} $AgNO_3$ = 0.00071 g of chloride)

7. Nitrate

Unlike carbonate, bicarbonate and many other anions, the presence of nitrate in irrigation waters is considered to be beneficial, particularly in slightly saline waters. Being an immediately available form of nitrogen, it is reported to benefit plants. However, extremely high nitrate content in water contaminated with various kinds of wastes including industrial effluents, may pollute the ground water, rendering it unfit for drinking. Therefore, for quality rating of irrigation water, nitrate determination is generally not included, except in cases where very high nitrate contents are suspected. Nitrate content in water can be determined by using phenoldisulphonic acid as done for soil extracts (See section 1.2.7). Alternatively, an autoanalyser, if available, can also be used which provides more precise results. The quantity of water sample (10 to 50 mL) to be used will depend upon the nitrate content, and the working standards may also have to be prepared accordingly.

8. Sulphate

The water sample has to be filtered using whatman No. 1 filter paper to remove physical impurities. If the sample is coloured, it should be decolourized with activated charcoal. Thereafter, turbidimetric procedure (Chesnin & Yien, 1950) a described in section 1.2. 12 can be adopted for determination of sulphate–S in water.

9. Boron

For irrigation quality rating, water samples are not usually tested for B content. However, saline waters of arid and semi-arid regions may have high concentrations of B, which could be toxic depending upon the type of minerals present on soil and the nature strata through which the water passes to reach the groundwater. Thus, some water from entrapped underground sources may contain excessive amount of boron, which is toxic to plants. In fact, the margin between toxicity and deficiency limits of boron is very narrow and hence, water samples suspected to be containing high B should be tested for B and classified accordingly. Boron in water can be determined by azomethine-H method, as described for soil boron under section 1.2.13.

c) Biochemical Oxygen Demand

The biochemical oxygen demand (BOD) of sewage or effluent indicates the amount of organic compounds present in it. It considered as a good index of organic pollution. High amount of organic matter in water leads to higher oxygen utilization by bacteria for degrading it. Use of sewage having high BOD increases the concentration of soluble organic compounds in the aquatic body, where it is discharged. Digestion of these compounds in natural ecosystems can deplete available oxygen, thus leading to asphyxiation of fish. Fortunately, most of the groundwater's and well or canal waters do not contain high organic matter.

Principle

The determination of BOD is based on the measurement of the volume of oxygen by bacteria to metabolize the organic compounds. The BOD is generally measured by incubating the water sample at 20°C for five days in the dark in aerobic conditions. In the tropical and sub-tropical climate, where both temperature and rate of metabolic activities are high, the incubation is preferably done at 27°C for three days. Since a significant amount of oxygen is required for nitrification, there is an over-estimation of BOD, if it is not checked by a suitable method, *e.g.* by adding 1 mL of 0.5% solution of allyl thiourea.

Instruments

1. pH meter
2. BOD incubator

Reagents

1. **Allyl thiourea:** Prepare 0.5% solution.
2. **Hydrochloric acid:** Prepare approximately 1 *N* solution.
3. **Sodium hydroxide:** Prepare approximately 1 N solution.
4. **Sodium thiosulphate:** Prepare 0.02 N aqueous solution (eq. wt. 248.18g)
5. Manganous sulphate solution
6. Alkaline iodide-azide solution
7. **Starch indicator:** Prepare 1% aqueous solution.
8. Conc. Sulphuric acid

Procedure

1. Neutralize the water sample using 1 N acid or 1 N alkali (attain a pH of 7.0).
2. Fill the water sample in 6 BOD bottles (250 mL) without bubbling.
3. Add 1 mL of allyl thiourea to each bottle.
4. Add 2 mL of manganous sulphate and alkaline iodide-azide solutions in succession to three of the six BOD bottles, right at the bottom with separate pipettes and replace the stopper.

5. Shake the bottles in the upside down direction at least six times.
6. Allow the brown precipitate to settle.
7. Add 2 mL of conc. H_2SO_4 and shake stoppered bottle to dissolve the brown precipitate.
8. Take 50 mL aliquot in a 250 mL conical flask and titrate with sodium thiosulphate solution till the colour changes to pale straw.
9. Add 2 drops of starch solution to the flask. It will change the colour from pale to blue.
10. Titrate again till the blue colour disappears.
11. Calculate the dissolved oxygen in mg L^{-1}, as given below.
12. Take the mean of the dissolved oxygen of the three bottle (D_1).
13. Incubate the rest three BOD bottles at 27°C in a BOD incubator for 3 days.
14. Estimate the oxygen concentration in all the three incubated samples and take mean (D_2).

Calculations

$$\text{Dissolved oxygen (mgL}^{-1}) = \frac{(8 \text{ X } 1000 \text{ X } N) \text{ X } t}{V}$$

Where, V = volume of sample taken,

T = total titre value (mL), and

N = normality of titrant (*i.e.* 0.025 *N*).

BOD (Mg L^{-1}) = D_1-D_2

Where, D_1 and D_2 are the dissolved oxygen contents in the initial (unincubated) and the incubated samples, respectively.

d) Chemical Oxygen Demand

With the alarming increase in pollution through the discharge of large amounts of various chemically oxidizable organic substances entering the aquatic systems, the biochemical oxygen demand alone does not give a precise estimate of the organic load of a water sample. Besides this, the presence of various toxicants in the sample may severely affect the reliability of BOD results. Therefore, the chemical oxygen demand (COD) is considered as a better indicator of organic matter.

Principle

The organic matter content of water is estimated by their oxidability by using chemical oxidants like potassium permanganate or potassium dichromate. In the case of the former, a know volume of $KMnO_4$ is first used to oxidize the organic matter and then the excess of oxygen is allowed to react with potassium iodide. This liberates iodine in amounts equal to the excess oxygen. The liberated iodine is estimated by sodium thiousulphate titration using starch as indicator.

Apparatus

Water bath

Reagents

1. 0.1N Potassium dichromate solution
2. 0.1 N Sodium thiosulphate solution
3. 2M Sulphuric acid
4. 1% Starch solution

Boron

1. Transfer 50 mL water sample in triplicate in 100 mL conical flask.
2. Carry distilled water blank also (in triplicate).
3. Add 5 mL of $K_2Cr_2O_7$ solution to each flask.
4. Place the flasks in water bath at 100°C for one hour.
5. Cool the flask for 10 minutes and add 5 mL of KI solution and 10 mL of H_2SO_4 to each flask.
6. Titrate the contents against 0.1 *M* sodium thiosulphate until plae yellow colour appears.
7. Add 1 mL of starch solution (it gives blue colour).
8. Titrate again until blue colour disappears completely.
9. Find out the total volume of titrant consumed and take the mean of the three flask of each sample.

10. Find out the total volume of the titrant used for the blank also.

Calculations

$$COD\ (mgL^{-1}) = \frac{8 X C X (B\text{-}A)}{V}$$

Where, C = concentration of titrant (m mol L^{-1}),

A = titre value for blank,

V = volume of water sample taken

e) Rating of irrigation water quality

Though all the determinations described for water samples in this chapter might be necessary in specific situations, the rating of water for irrigation is usually carried out on the basis of the not-so-specific tests. These are salinity, alkalinity, sodium hazard and Ca+ Mg only, referring to the chart. The chart prepared by the Soil Testing Laboratory,IARI, New Delhi is based on the relationship of the Ca+Mg and electrical conductivity values with the degree of sodium hazard. For using this chart, adopt the following method:

1. Locate the determined value of Ca+ Mg in me L^{-1} of the water sample on the upper scale of the X-axis.
2. Draw a perpendicular from this point.
3. Now locate the value of electrical conductivity (dS m^{-1}) on the lower scale of X-axis and draw a slanting line from this point parallel to the slanting lines provided in the chart.
4. Find out the point where this line intersects the perpendicular. This point lies in specific zone with thick-line boundaries which gives the water class name. this class name has numeral values with 'C' and 'S' which stand for conductivity and sodium, respectively.

Note: The approximate value of SAR can also be obtained from this chart by drawing a line from the point of intersection parallel to X-axis, to meet the SAR line on Y-axis. Once the water class name is obtained from the chart, work out the water suitability for irrigation following the procedure given below:

1. Make a total of the numeral values of 'C' and 'S' from the water class obtained (Total-I).
2. Pick up the weightage value for the textural class of the soil to be irrigated from the following table and add this to the total-I to get total-II :

Soil Textural Class	Weightage
Sandy	2.0
Sandy loam	2.5
Loam	3.0
Clay loam	3.5
Clay	4.0

3. Subtract total-II form '9' to obtain a difference (D), which will indicate the suitability for irrigating the number of categories of crops, priority starting from salt tolerant crops.
4. Recommend the water for all the three crop categories given in Table 4 (*i.e.* sensitive, semi-salt tolerant) if D=3, for the second and third categories if D=2, and for only the tolerant crops, if D=1.
5. In case the value of D = 2.5, accept the water for irrigating sensitive crops also but with dilution and/or amendment.
6. If D = 1.5, recommend for amendment.

f) Use of saline/sodic waters for irrigation

As described above, water can be safely used fro irrigation as long as the sum total of all the indices and weightage for soil texture does not exceed 9. Highly saline water, however, can not be recommended for all crop categories either due to high value of residual sodium carbonate of high sodium adsorption ratio. If another source of good quality irrigation water is available, the high SAR water can be used after diluting it suitably. Dilution does not strictly mean the mixing of the two sources in a particular ratio but aim could be achieved by alternate irrigations as well. Dilution in more than 1:2 ratio is generally not recommended obviously because such high SAR waters may as such be dropped and the better source alone used for irrigation. In case the availability of good water is very limited, the saline water may be used at a later stage of plant growth while the germination and early growth stages may be managed with safe water.

Waters having high amounts of residual sodium carbonate (RSC) may also be used, but along with the required quantity of gypsum to neutralize the excess RSC. For all practical purposes, gypsum may be recommended for soil application @ 40 kg/irrigation ha for each milliequivalent of RSC to be neutralized.

g) Interpretation of water analysis data

Various workers have suggested the criteria of classification of irrigation waters on the basis of analysis made (Richards, 1954; Paliwal and Yadav, 1976; Ayers Westcot, 1976). Some important and widely acceptable ratings are given below. These should be taken as a general guideline and necessary correction may be made depending upon the soil-crop situation.

Parameter	Permissible	Moderately safe	Moderately unsafe	Unsafe
RSC (me L^{-1})	< 1.25	-	1.25-2.50	> 2.50
SAR	< 10	10-18	18-26	> 26
Boron (mg L^{-1})	< 2.0	2.0-2.5	2.5-3.0	> 3.0
Chloride (mg L^{-1})	< 140	140-350	-	> 350
BOD (mg L^{-1})	d" 100	-	-	> 100

Table 3.1: Salt tolerance of crops

Category	Crops
Sensitive	
Field crops	Field bean, green gram lentil
Forage crops	Red clover, white clover, *guar* (cluster bean)
Vegetables	Celery, radish (English var.), bean
Fruits	Peach, apricot, pear, apple, plum, strawberry, blackberry
Semi-salt tolerant	
Field crops	Oats, rice, sorghum, maize, pearl millet, wheat (improved var.), pigeonpea, gram caster, flax, soybean, rye, caster bean
Forage crops	Senji (Melilotus sp.) menthe, sorghum, maize, berseem, cowpea, ryegrass, sudan grass, oat

Vegetables	Tomato, cabbage, cauliflower, lettuce, potato, radish, carrot, onion, lady finger, pea, cucumber, pumpkin, sweet potato, squash
Fruits	Grape, olive, fig, guave, mango, banana, pomegranate, orange, grape fruit, lemon, almond, pineapple
Salt tolerant	
Field crops	Barley, *dhaincha* (Sesbania sp) sugarbeet, tobacco, cotton, wheat (some local var.), sugarcane, rapeseed
Forage crops	Salt grass, dubgras (Cyanodon sp.), Rhodes grass, Bermuda grass
Vegetables	Beetroot, asparagus, spinach, kale, turnip
Fruits	Date plam, coconut, falsa (*Grewia sp.*)

❑❑❑

References

Bear, F.E. and Mc Clure, G.M. (1920). Sampling soil plots. Soil Sci., 9: 65-75.

Belchev, D.J. (1952). The measurement of soil moisture and density by neutron and gamma ray scattering. Highway Res. Board Spl. Rpt. 2: 98-110.

Bouyoucos, G.J. (1937). Evaporation water with burning alcohol as a rapid means of determining moisture content of soils. Soil Sci. 44: 337-383.

Boyadgiev, T. (1985). Salty soils of arid and semi-arid regions Sci. Rev. Arid Zones Res., Vol. 3:1-47. Scientific Publisher, Jodhpur.

Brady, N.C. (1990). The Nature and Properties of Soils: Macmillan Publ. Co. New York, NY.

Bragg, W.L. (1933). The crystalline state. Macmillan Publ. Co., New York, NY, USA

Brindley, G.W. (1980). Quantitative x-ray mineral analysis of clays. pp. 141-438. In: Crystal structures of clays Minerals and their x-ray identification. Mineral. Soc. monograph No. 5. Mineralogical Society, London, England.

Bureau, P. and Roederer, P. (1960). Contribution a l'etude des sols gypseux de la partie sud du golfe de Gabes. Bull. Assoc. France Etude Sol, 8: 145-149. Section spec. d'Etudes de pedologie et d'Hydrologie, Es 33. 107.

Carson, C.D. and Dixon, J.B. (1972). Potassium selectivity in certain montmorillonitic soil clays. Soil Sci. Soc. Am. Proc., 36: 838-843.

Carter, M.R. (1981). Association of total $CaCO_3$ and active $CaCO_3$ with growth of fine tree species on chernozemic soils. Canadian J. Soil Sci., 61:173- 175.

Chang, F.H. (1974). Quantitative interpretation of x-ray diffraction patterns of mixtures. I. Matrix-flushing method for quantitative multi-component analysis. J. Appl. Cryst., 7: 519-525.

Cline, M.G. (1944). Principles of soil sampling. Soil Sci., 58: 275-288.

Coghill, W.H. and Devaney, F.D. (1937). Ball-mill grinding. U.S. Bureau Mines. Techn. Paper No. 581.

Coutinet, S. (1965). Methodes d'analyse utilisables pour les sols sales, calcaires et gypseux. Agronomie Tropicale, Paris, 12: 1242-1253.

Deb, B.G. and Chadha, S.P. (1970). Mechanical analysis of calcareous soils and distribution of CaCO3 in various fractions. Journal of Indian Society of Soil Science, 18: 227-232.

Harward, M.E., Carstea, D.D. and Sayegh, A.H. (1969). Properties of vermiculites and smectites: expansion and collapse. Clays and Clay Minerals, 16: 437-447.

Hemwall, J.B. (1957). The fixation of phosphorus by soils. Advances in Agronomy 1957, pp. 95-112.

Hesse, P.R. (1972). A Textbook of Soil Chemical Analysis. Chem. Publ. Co., New York, NY.

Hesse, P.R. (1974). Methods of soil analysis-texture analysis of gypsic soils. The Hurlbut (Jr.), C.S. and Klein, C. 1977. Manual of Mineralogy. John Wiley & Sons, New York, NY.

Jackson, M.L. (1956). Soil Chemical Analysis - Advanced Course. Dept. Soils, Univ. of Wisconsin, Madison, WI.

Keeney, D.R. and Nelson, D.W. (1982). Nitrogen - inorganic forms. pp. 643-698. In: Methods of Soil Analysis, Part 2, A.L.

Kim, H.T. (1996). Soil Sampling, preparation and analysis. Marcel Dekker Inc., New York, NY, 10016. Madison, WI.

Lindsay, W.L. (1973). Inorganic phase equilibria of micronutrients in soils. pp. 41-57. In: Micronutrients in Agriculture. J.J. Mordvedt, P.M. Giordano, and W.L. Lindsay (eds). Soil Sci. Soc. Am. Inc., Madison, WI.

Lindsay, W.L. and Norvell, W.A. (1978). Development of DTPA soil test for zinc, iron, manganese and copper. Soil Sci. Soc. Am. J., 42: 421-428.

Marwan, H.K., Sayegh, A.H., and Atallah, N.J. (1978). The effect of soil conditioners on pore size distribution and water retention of a calcareous soil. Aust. J. Soil Res., 29: 277-81.

Matar, A.E. and Douleimy, T. (1978). Note on a proposed method for the mechanical analyses of gypsiferous soils. ACSAD publication. The

Arab Center for the Studies of Arid Zones and Dry Lands, Damascus, Syria.

Norvell, W.A. (1973). Equilibria of metal chelates in soil solution. pp. 115-138. In: Micronutrient in Agriculture, J.J. Mortvedt, P.M. Giordano, and W.L. Lindsay (eds). Soil Sci. Soc. Am. Inc., Madison, WI, USA. 109

Petersen, R.G. and Chamblee, D.S. (1955). Optimum size of sample for hand separation of forage crop mixtures into their component species in small plot experiments. Agron. J., 47: 20-23.

Pierzynski, G.M., Sims, J.T. and Vance, G.F. (2000). Soils & Environmental.

Richards, L.A. (ed.). (1954). United State Salinity Laboratory Staff, Handbook No. 60. Diagnosis and improvement of Saline and Sodic Soils. United State Department of Agriculture, U.S. Government Printing Office, Washington D.C., USA.

Sayegh, A.H., Khan, N.A., Khan, P. and Ryan, J. (1978). Factors affecting gypsum and cation exchange capacity determinations in gypsiferous soils. Soil Sci., 125(5): 194-300.

Sayegh, A.H. (1986). In: Management of gypsiferous soil. FAO Soils Bulletin 62, Rome, Italy.

Schoonover, (1953). Examination of soils for alkali-quick tests. University of Calif. Ext. Service, Berkeley, Calif (mimeo).

Soltanpour, P.N., Olsen S.R. and Goos, R.J. (1982b). Effect of nitrogen fertilization on dryland wheat on grain selenium concentration. SSSAJ., 46: 430-433.

Sparks, D.L. (ed). (1996). Methods of Soil Analysis, Part three, Chemical Methods. ASA, Madison, Wisconsin, USA.

Sys, C. (1975). Land evaluation in Europe. FAO Soils Bulletin 29, pp. 113, Rome, Italy.

Tan, K.H. (1994). Environmental Soil Science. Marcel Dekker Inc., New York, NY.

Tan, K.H. (1995). Soil Sampling, Preparation, and Analysis. Marcel Dekker Inc., New York, N.Y. USA.

Tarzi, G.J. (1984). Procedures for collecting soil samples for different purposes.

Ministry of Agriculture and Water, Directorate of Agricultural Research, Riyadh, Kingdom of Saudi Arabia.

Thabet, A. (1975). Active and total $CaCO_3$ in soils. Thesis. American University of Beirut, Beirut, Lebanon.

Thomson, A.P. Duthie, D.M.L. and Wilson, M.J. (1972). Randomly oriented powders for quantitative identification by x-ray diffraction. Clay Mineral, 9: 345-348.

Tisdale, S.L. and Nelson, W.L. (1993). Soil Fertility and Fertilizers. Macmillan Co., New York, N.Y.

Van Alphen, J.G. and Rios Romero, F. (1971). Gypsiferous soils. Bulletin 12, International Institute for Land Reclamation and Improvement. Wageningen, the Netherlands.

Van Schuylenborgh, J. (1954). The effect of air-drying of soil samples upon some physical soil properties. Neth. J. Agri. Sci., 2: 50-57.

Vieillefon, J. (1979). Contribution to the improvement of analysis of gypsiferous soils. Cahiers/ORSTOM, Serie Pedologie, 17: 195-223.

Walkley, A. (1947). A critical examination of a rapid method for determining organic carbon in soils. Effect of variations in digestion conditions and inorganic soil constituents. Soil Sci., 63: 251-263.

Watanabe, F.S. and Olsen, S.R. (1965). Test of an ascorbic acid method for determining phosphorus in water and NaHCO3 extracts from soil. Soil Sci. Am. Proc., 29: 677-678.

Waynick, D.D. (1918). Variability in soils and its significance to past and future soil investigations. Univ. of California Pub. in Agr. Sci., 3: 243-269.

Westin F.C. and Buntley, G.J. (1966). Soil phosphorus in South Dakota I: inorganic phosphorus fixation of some soil series. Soil Sci. Soc. Amer. Proc., 30: 245.

Whittig, L.D. and Allardice, W.R. (1986). X-ray diffraction techniques. pp.331-362. In: Methods of Soil Analysis, Part 1, A. Klute (ed.). Agronomy series No.

Belchev, D.J. (1952). The measurement of soil moisture and density by neutron and gamma ray scattering. Highway Res. Board Spl. Rpt. 2: 98-110.

Bouyoucos, G.J. (1937). Evaporation water with burning alcohol as a rapid means of determining moisture content of soils. Soil Sci. 44: 337-383.

VanBavel, C.H.M. (1958). Measurement of soil moisture content by the neutron method. Agr. Res. Serv. ARS. 41-24.

□□□

Appendix

Appendix A. List of the atomic weights of the elements

Element	Symbol	Atomic Number	Atomic Weight	Element	Symbol	Atomic Number	Atomic Weight
Actinium	Ac	89	(227)	Mercury	Hg	80	200.59
Aluminium	Al	13	26.98	Molybdenum	Mo	42	95.94
Amerivium	Am	95	(243)	Neodymium	Nd	60	144.24
Antimony	Sb	51	121.75	Neon	Ne	10	20.183
Argon	Ar	18	39.948	Neptunium	Np	93	(237)
Arsenic	As	33	74.92	Nickel	Ni	28	58.71
Astatine	At	85	(210)	Niobium	Nb	41	92.91
Barium	Ba	56	137.34	Nitrogen	N	7	14.007
Berkelium	Bk	97	(249)	Nobelium	No	102	(253)
Beryllium	Be	4	9.012	Osmium	Os	76	190.2
Bismuth	Bi	83	208.98	Oxygen	O	8	15.9994
Boron	B	5	10.81	Palladium	Pd	46	106.4
Bromine	Br	35	79.909	Phosphorus	P	15	30.974
Cadmium	Cd	48	112.40	Platinum	Pt	78	195.09
Calcium	Ca	20	40.08	Plutonium	Pu	94	(242)
Californium	Cf	98	(251)	Polonium	Po	84	(210)
Carbon	C	6	12.011	Potassium	K	19	39.102
Cerium	Ce	58	140.12	Praseodymium	Pr	59	140.91
Cesium	Cs	55	132.91	Promethium	Pm	61	(147)
Chlorine	Cl	17	35.453	Protactinium	Pa	91	(231)
Chromium	Cr	24	52.00	Radium	Ra	88	(226)
Cobalt	Co	27	58.93	Radon	Rn	86	(222)
Copper	Cu	29	63.54	Rhenium	Re	75	186.23
Curium	Cm	96	(247)	Rhodium	Rh	45	102.91
Dysprosium	Cy	66	162.50	Rubidium	Rb	37	85.47
Einsteinium	Es	99	(254)	Ruthenim	Ru	44	101.1
Erbium	Er	68	167.26	Samarium	Sm	62	150.35

Europium	Eu	63	151.96	Scandium	Sc	21	44.96
Fermium	Fm	100	(253)	Selenium	Se	34	78.96
Fluorine	F	9	19.00	Silicon	Si	14	28.09
Francium	Fr	87	(223)	Silver	Ag	47	107.870
Gadolinium	Gd	64	157.25	Sodium	Na	11	22.9898
Gallium	Ga	31	69.72	Strontium	Sr	38	87.62
Germanium	Ge	32	72.59	Sulfur	S	16	32.064
Gold	Au	79	196.97	Tantalum	Ta	73	180.95
Hafnium	Hf	72	178.49	Technetium	Tc	43	(99)
Helium	He	2	4.003	Tellurium	Te	52	127.60
Holmium	Ho	67	164.93	Terbium	Tb	65	158.92
Hydrogen	H	1	1.0080	Thallium	TI	81	204.37
Indium	In	49	114.82	Thorium	Th	90	232.04
Iodine	I	53	126.90	Thulium	Tm	69	168.93
Iridium	Ir	77	192.2	Tin	Sn	50	118.69
Iron	Fe	26	55.85	Titanium	Ti	22	47.90
Krypton	Kr	36	83.80	Tungsten	W	74	183.85
Lanthanum	La	57	138.91	Uranium	U	92	238.03
Lawrencium	Lw	103	(257)	Vanadium	V	23	50.94
Lead	Pb	82	207.19	Xenon	Xe	54	131.30
Lithium	Li	3	6:939	Ytterbium	Yb	70	173.04
Lutetium	Lu	71	174.97	Yttrium	Y	39	88.91
Magnesium	Mg	12	24.312	Zinc	Zn	30	65.37
Manganese	Mn	25	54.94	Zirconium	Zr	40	91.22
Mendelevium	Md	101	(256)				

Number in parentheses indicate mass of most stable known isotopes.

Appendix B. pH indicators

pH Indicators		pH transition intervals		
Name	**Colour**	**pH**	**pH**	**Colour**
Cresol red	Pink	0.2	1.8	Yellow
m-Cresol purple	Red	1.2	2.8	Yellow
Thymol blue	Red	1.2	2.8	Yellow
p-Xylenol blue	Red	1.2	2.8	Yellow
Triphenylcarbinol	Red	1.2	3.2	Colourless
2,4-Dinitrophenol	Colourless	2.8	4.7	Yellow
4-Dimethylaminoa-zobenzene	Red	2.9	4.0	Yellow-Orange
Bromochlorophenol blue	Yellow	3.0	4.6	Purple
Bromophenol blue	Yellow	3.0	4.6	Purple
Methyl orange	Red	3.1	4.4	Yellow-Orange

Bromocresol green	Yellow	3.8	5.4	Blue
2,5-Dnitrophenol	Colourless	4.0	5.8	Yellow
Alizarinsulfonic acid sodium salt	Yellow	4.3	6.3	Violet
Methyl red	Red	4.4	6.2	Yellow-Orange
Methyl red sodium salt	Red	4.4	6.2	Yellow-Orange
Chlorophenol red	Yellow	4.8	6.4	Purple
Hematoxylin	Yellow	5.0	7.2	Violet
Litmus extra pure	Red	5.0	8.0	Blue
Bromophenol red	Orange-yellow	5.2	6.8	Purple
Bromocresol purple	Yellow	5.2	6.8	Purple
4-Nitrophenol	Colourless	5.4	7.5	Yellow
Bromoxylenol blue	Yellow	5.7	7.4	Blue
Alizarin	Yellow	5.8	7.2	Red
Bromothymol blue	Yellow	6.0	7.6	Blue
Phenol red	Yellow	6.4	8.2	Red
3-Nitrophenol	Colourless	6.6	8.6	Yellow-Orange
Neutral red	Bluish-red	6.8	8.0	Orange- Yellow
4,5,6,7-Tetrabromo-phenolphthalein	Colourless	7.0	8.0	Purple
Cresol red	Orange	7.0	8.8	Purple
1-Naphtholphthalein	Brownish	7.1	8.3	Blue-Green
m-Cresol purple	Yellow	7.4	9.0	Purple
Thymol blue	Yellow	8.0	9.6	Blue
P-Xylenol blue	Yellow	8.0	9.6	Blue
Phenolphthalein	Colourless	8.2	9.8	Red-Violet
Thymolphthalein	Colourless	9.3	10.5	Blue
Alizarin yellow GG	Light yellow	10.2	12.1	Brownish-Yellow
Epsilon blue	Orange	11.6	13.0	Violet

Appendix C. Concentrations of Commercial Reagent-Grade Acids and bases[a]

Reagent	F-Wt.[b]	M[c]	% by Wt.	Density (20⁰), g/cm³
H_2SO_4	98.08	17.6	94.0	1.831
$HClO_4$	100.5	11.6	70.0	1.668
HCl	36.46	12.4	38.0	1.188
HNO_3	63.01	15.4	69.0	1.409
H_3PO_4	98.00	14.7	85.0	1.686
$HC_2H_3O_2$	60.05	17.4	99.5	1.051
NH_3	17.03	14.8	28.0	0.898

[a]These are approximate concentrations and cannot be used for preparing standard solutions. [b]Formula weight. [c]Molarity.

These are approx concentration and cannot be used for various standard solution of formula weight Molarity

Neon	Ne	10	20.183
Argon	Ar	18	39.948
Neptunium	Np	93	237

Appendix D. Temperature factors (f_t) for correcting resistance and conductivity data on soil extracts to the standard temperature of 25^{0}C

			$EC_{25} = EC_t \times f_t; EC_{25} = (k/R_t) \times f_t; R_{25} = R_t/f_t$					
^{0}C	^{0}F.	f_t	^{0}C	^{0}F.	f_t	^{0}C	^{0}F.	f_t
3.0	37.4	1.709	22.0	71.6	1.064	29.0	84.2	0.925
4.0	39.2	1.660	22.2	72.0	1.060	29.2	84.6	0.921
5.0	41.0	1.613	22.4	72.3	1.055	29.4	84.9	0.918
6.0	42.8	1.569	22.6	72.7	1.051	29.6	85.3	0.914
7.0	44.6	1.528	22.8	73.0	1.047	29.8	85.6	0.911
8.0	46.4	1.488	23.0	73.4	1.043	30.0	86.0	0.907
9.0	48.2	1.488	23.2	73.8	1.038	30.2	86.4	0.904
10.0	50.0	1.411	23.4	74.1	1.034	30.4	86.7	0.901
11.0	51.8	1.375	23.6	74.5	1.029	30.6	87.1	0.897
12.0	53.6	1.341	23.8	74.8	1.025	30.8	87.4	0.894
13.0	55.4	1.309	24.0	75.2	1.020	31.0	87.8	0.890
14.0	57.2	1.277	24.2	75.6	1.016	31.2	88.2	0.887
15.0	59.0	1.247	24.4	75.9	1.012	31.4	88.5	0.884
16.0	60.8	1.218	24.6	76.3	1.008	31.6	88.9	0.880
17.0	62.6	1.189	24.8	76.6	1.004	31.8	89.2	0.877
18.0	64.4	1.163	25.0	77.0	1.000	32.0	89.6	0.873
18.2	64.8	1.157	25.2	77.4	0.996	32.2	90.0	0.870
18.4	65.1	1.152	25.4	77.7	0.992	32.4	90.3	0.867
18.6	65.5	1.147	25.6	78.1	0.988	32.6	90.7	0.864
18.8	65.8	1.142	25.8	78.5	0.983	32.8	91.0	0.861
19.0	66.2	1.136	26.0	78.8	0.979	33.0	91.4	0.858
19.2	66.6	1.131	26.2	79.2	0.975	34.0	93.2	0.843
19.4	66.9	1.127	26.4	79.5	0.971	35.0	95.0	0.829
19.6	67.3	1.122	26.6	79.9	0.967	36.0	96.8	0.815
19.8	67.6	1.117	26.8	80.2	0.964	37.0	98.6	0.801
20.0	68.0	1.112	27.0	80.6	0.960	38.0	100.2	0.788
20.2	68.4	1.107	27.2	81.0	0.956	39.0	102.2	0.775
20.4	68.7	1.102	27.4	81.3	0.953	40.0	104.0	0.763
20.6	69.1	1.097	27.6	81.7	0.950	41.0	105.8	0.750
20.8	69.4	1.092	27.8	82.0	0.947	42.0	107.6	0.739
21.0	69.8	1.087	28.0	82.4	0.943	43.0	109.4	0.727
21.2	70.2	1.082	28.2	82.8	0.940	44.0	111.2	0.716
21.4	70.5	1.078	28.4	83.1	0.936	45.0	113.0	0.705
21.6	70.9	1.073	28.6	83.5	0.932	46.0	114.8	0.694
20.8	71.2	1.068	28.8	83.8	0.929	47.0	116.6	0.683

Source: Agricultural Handbook 60, U.S. Deptt. of Agriculture

Appendix E. Conversion factors for SI and non-SI units (Soil Science Society of America Journal)

To convert Column 1 into Column 2, multiply by	Column 1 SI Unit	Column 2 Non-SI Unit	To convert Column 2 into Column 1 multiply by
		Length	
0.621	Kilometre, km (10^3 m)	Mile, mi	1.609
1.094	Metre, m	Yard, yd	0.914
3.28	Metre, m	Foot, ft	0.304
1.0	Micrometre, µm (10^{-6} m)	Micron, µ	1.0
3.94×10^{-2}	Millimetre, mm (10^{-3} m)	Inch, in	25.4
10	Nanometre, nm (19^{-9} m)	Angstrom, a	0.1
		Area	
2.47	Hectare, ha	Acre	0.405
247	Square kilometre, km^2 (10^{-3} m)^2	Acre	4.05×10^{-3}
0.386	Square kilometre, km^2 (10^{-3} m)^2	Square mile, nil^2	2.590
2.47×10^{-4}	Square metre, m^2	Acre	4.05×10^{3}
10.76	Square metre, m^2	square foot, ft^2	9.29×10^{-2}
1.55×10^{-3}	Square millimetre, mm^2 (10^{-3} m)^2	Square inch, in^2	645
		Volume	
9.73×10^{-3}	Cubic metre, m^3	Acre-inch	102.8
35.3	Cubic metre, m^3	Cubic foot, ft^3	2.83×10^{3}
6.10×10^{4}	Cubic metre, m^3	Cubic inch, in^3	1.64×10^{-5}
2.84×10^{-2}	Liter, L (10^{-3} m^3)	Bushel, bu	35.24
1.057	Liter, L (10^{-3} m^3)	Quart (liquid) qt	0.946
3.53×10^{-2}	Liter, L (10^{-3} m^3)	Cubic foot, ft^3	28.3
0.265	Liter, L (10^{-3} m^3)	Gallon	3.78
33.78	Liter, L (10^{-3} m^3)	Pint (tluid), pt	0.473

Mass			
2.20×10^{-3}	Gram, g (10^{-3} kg)	Pound, lb	454
3.52×10^{-2}	Gram, g (10^{-3} kg)	Ounce (avdp), oz	28.4
2.205	Kilogram, kg	Pound, lb	0.454
0.01	Kilogram, kg	Quintal (metric), q	100
1.10×10^{-3}	Kilogram, kg	ton (2000 lb), ton	907
1.102	Megagram, Mg (tonne)	Ton (U.S.), ton	0.907
1.102	Tonne, t	Ton (U.S.), ton	0.907
Yield and Rate			
0.893	Kilogram per hectare, kg ha^{-1}	Pound per acre, lb $acre^{-1}$	1.12
7.77×10^{-2}	Kilogram per cubic meter, kg m^{-3}	Pound per bushel, lb bu^{-1}	12.87
1.49×10^{-2}	Kilogram per hectare, kg ha^{-1}	Bushel per acre, 60 lb	67.19
1.59×10^{-2}	Kilogram per hectare, kg ha^{-2}	Bushel per acre, 56 lb	62.71
1.86×10^{-2}	Kilogram per hectare, kg ha^{-1}	Bushel per acre, 48 lb	53.75
0.107	Liter per hectare, L ha^{-1}	Gallon per acre	9.35
893	Tonnes per hectare, Mg ha^{-1}	Pound per acre, lb $acre^{-1}$	1.12×10^{-3}
893	Megagram per hectare, Mg ha^{-1}	Pound per acre, lb $acre^{-1}$	1.12×10^{-3}
0.446	Megagram per hectare, Mg ha^{-1}	Ton (2000 lb) per acre, ton $acre^{-1}$	2.24
2.24	Meter per second, m s^{-1}	Mile per hour	0.447

Appendix E. (*continued*)

To convert Column 1 into Column 2, Multiply by	Column 1 SI Unit	Column 2 Non-SI Unit	To convert Column 2 into Column 1 Multiply by
	Specific Surface		
10	Square metre per kilogram, m^2 kg^{-1}	Square centimetre per gram cm^2 g^{-1}	0.1
1000	Square metre per kilogram, m2 kg-1	Square millimetre per gram, mm^2g^{-1}	0.001
	Pressure		
9.90	Megapascal, MPa (10^6 Pa)	Atmosphere	0.101
10	Megapascal, MPa (10^6 Pa)	Bar	0.1
1.00	Megagram per cubic meter, Mg m^{-3}	Gram per cubic centimeter, g cm^{-3}	1.00
2.09×10^{-2}	Pascal, Pa	Pound per square foot, lb ft^{-2}	47.9
1.45×10^{-4}	Pascal, Pa	Pound per square inch, lb in^{-2}	6.90×10^3
	Temperature		
1.00 (K-273)	Kelvin, K	Celsius, ^{0}C	1.00 (^{o}C+273)
(9/5 ^{o}C)+32	Celsius, ^{o}C	Fahrenheit, ^{o}F	5/9 (^{o}F-32)
	Plane Angle		
57.3	Radian, rad	Degress (angle), ^{o}	1.75×10^{-3}
	Electrical Conductivity, Electricity, and Magnetism		
10	Siemen per metre, S m^{-1}	Millimho per centimetre, mmho cm^{-1}	0.1
10^4	Tesla, T	Gauss, G	10^{-4}
	Water Measurement		
9.73×10^{-3}	Cubic metre, m^3	Acre-inches, acre-in	102.8
9.81×10^{-3}	Cubic metre per hour, m^3h^{-1}	cubic feet per second, ft3 s^{-1}	101.9
4.40	Cubic metre per hour, m^3 h^{-1}	U.S. gallons per minutes, gal min^{-1}	0.227

8.11	Hectare-metre, ha-m	Acre-feet, acre-ft	0.123
97.28	Hectare-metre, ha-m	Acre-inches, acre-in	1.03×10^{-2}
8.1×10^{-2}	Hectare-centimetres, ha-cm	Acre-feet, acre-ft	12.33
Concentrations			
1	Centimole per kilogram, cmol kg-1 (ion exchange capacity)	Milliwquivalents per 100 grams, meq 100 g^{-1}	1
0.1	Gram per kilogram, g kg^{-1}	Percent, %	10
1	Milligram per kilogram, mg kg^{-1}	Parts per million, pom	1
Radioactivity			
2.7×10^{-11}	Becquerel, Bq	Curie, Ci	3.7×10^{10}
2.7×10^{-2}	Becquerel per kilogram, Bq kg^{-1}	Picocurie per gram, $_p$Ci g^{-1}	37
100	Grey, Gy (absorbed dose)	rad, rd	0.01
100	sievert, sv (equivalent dose)	Rem (roentgen equivalent man)	0.01
Plant Nutrient Conversion			
	Element	Oxide	
2.29	P	P_2O_5	0.437
1.20	K	K_2O	0.830
1.39	Ca	CaO	0.715
1.66	Mg	MgO	0.602

Appendix F. Electrical conductivity values at different $CaSO_4$ concentration in water

$CaSO_4$ Conc. (meq/l)	EC at 250CMs/cm
1.0	0.121
2.0	0.226
5.0	0.500
10.0	0.900
20.0	1.584
30.0	2.205

Source: Richards, 1954

Appendix G. Plant Nutrient conversion

Element	Oxide
2.29 P	P_2O_5 0.437
1.20 K	K_2O 0.830
1.39 Ca	CaO 0.715
1.66 Mg	MgO 0.602

□□□